Characterization of Polymers in Solution

Yo Nakamura · Ken Terao · Tomohiro Hirano ·
Atsushi Takano

Characterization of Polymers in Solution

Experimental Determination of Molecular Parameters

Springer

Yo Nakamura
Kyoto University
Kyoto, Japan

Ken Terao
The University of Osaka
Toyonaka, Japan

Tomohiro Hirano
Tokushima University
Tokushima, Japan

Atsushi Takano
Nagoya University
Nagoya, Japan

ISBN 978-981-96-6676-8 ISBN 978-981-96-6677-5 (eBook)
https://doi.org/10.1007/978-981-96-6677-5

This Springer imprint is published by the registered company Springer Nature Singapore Pte Ltd.
The registered company address is: 152 Beach Road, #21-01/04 Gateway East, Singapore 189721, Singapore

Preface

When researchers unfamiliar with polymer science begin their research on polymer materials, the first barrier they face may be the characterization of the polymers of interest. To find a way to do that, they may try to look for a book or papers about the measurements. However, this process may be time-consuming, since some texts go deep into the principles of the measurement and some papers only deal with very specific systems. A concise book that describes the experimental techniques may be essential to start the experiment quickly. This book focuses on experimental techniques and data analysis methods, omitting most of the measurement principles. In addition, the chapters and sections are categorized by the quantities obtained, which may allow the experimenters quickly to access the information they need. The authors hope that this book will be useful for university students, young scientists in research facilities, or researchers in companies who are not familiar with this field.

Kyoto, Japan
December 2024

On behalf of the authors,
Yo Nakamura

Contents

1 Introduction 1
  1.1 Background 1
  1.2 What is Characterization? 3
  References 3

2 Molecular Weight and Molecular Weight Distribution 5
  2.1 Overview 5
    2.1.1 Average Molecular Weights 5
    2.1.2 Polydispersity Index 7
  2.2 Molecular Weight 7
    2.2.1 Osmotic Pressure 7
    2.2.2 Vapor Pressure Osmometry 10
    2.2.3 Sedimentation Equilibrium 11
    2.2.4 Light Scattering 14
    2.2.5 Intrinsic Viscosity 17
  2.3 Molecular Weight Distribution 20
    2.3.1 Size-Exclusion Chromatography 20
    2.3.2 Mass Spectroscopy 23
    2.3.3 Comparison of Molecular Weight Distributions Determined by Different Methods 25
  References 28

3 Molecular Size and Shape 31
  3.1 Overview 31
  3.2 Radius of Gyration and Scattering Function 32
    3.2.1 Light Scattering 32
    3.2.2 Small-Angle X-ray and Neutron Scattering 36
    3.2.3 Measurements and Data Analysis 38

3.3 Hydrodynamic Properties 48
3.3.1 Dynamic Light Scattering 48
3.3.2 Sedimentation Velocity 48
3.3.3 Hydrodynamic Radii 51
3.4 Determination of Molecular Parameters 53
3.4.1 Wormlike Chain Parameters 53
3.4.2 Analysis of Long Chain Branching 60
3.4.3 Analysis of Aggregates 64
Appendix 66
References 67

4 Microstructures 71
4.1 Overview 71
4.2 Nuclear Magnetic Resonance 72
4.2.1 Stereochemical Analysis by NMR 72
4.2.2 Stereosequence Analysis with Polymerization Statistics 78
4.2.3 Monomer Sequence Analysis by NMR 84
4.2.4 Multivariate Analysis of NMR Spectra 91
4.2.5 Chain-End Analysis by NMR 93
4.3 Chromatographic Techniques 98
4.3.1 HPLC for Characterization of Polymers 98
4.3.2 Device Configuration of HPLC 98
4.3.3 Application of HPLC to Polymer Analysis: SEC 99
4.3.4 Three Types of HPLC Elution Modes Used for Polymer Analysis: SEC, IC, LCCC 100
4.3.5 Determination of Purity of Cyclic Polymers Using LCCC 101
4.3.6 Determination of Purity of Block Copolymers Using TGIC 103
References 106

Index 111

Chapter 1
Introduction

1.1 Background

It has been 100 years since human beings recognized polymers, which consist of many repeating units connected to make chain molecules. The pioneer Herman Staudinger proved the existence of such molecules by measuring molecular weights [1], mostly by the osmotic pressure measurement which was, at that time, the most reliable method to obtain molecular weight values of large molecules. Later, this rather time-consuming and less accurate method was replaced by more sophisticated techniques.

The accuracy of the molecular weight values has been improved by instrumental technologies, light scattering, and ultracentrifugation. The former technique was developed by Debye [2] and Zimm [3], and the latter by Svedberg [4]. They have established the theoretical background and designed the instruments. The advantage of these methods is that they give not only the molecular weight of the polymer, but also the dimension of the molecule, or the mean-square radius of gyration, by the former, and the hydrodynamic radius by the latter. The molecular weight dependence of the mean-square radius of gyration of the polymer has been shown to provide important information about the unperturbed dimension of the polymer coil in the theta solvents, and the magnitude of expansion of the coil by the repulsive intramolecular interaction, or the excluded-volume effect, in good solvents. Precise descriptions of these methods are described in Paul J. Flory's book "Principles of Polymer Chemistry" [5] and also in Charles Tanford's book "Physical Chemistry of Macromolecules" [6]. Studies on the molecular weight dependence of the dimensions of polymers were extended to the determination of the stiffness parameters of polymer chains [7, 8] and the branched structures of polymers [9]. With the advent of personal computers and laser beam sources, dynamic light-scattering instruments have been developed to determine the diffusion constant or the hydrodynamic radius of polymer molecules in solution [10].

Author: Y. Nakamura (nakamura.yo.7v@kyoto-u.ac.jp)

Y. Nakamura et al., *Characterization of Polymers in Solution*,
https://doi.org/10.1007/978-981-96-6677-5_1

Scattering techniques have been extended to other beam sources with short wavelengths, X-ray and neutrons. The former source is used to determine the radius of gyration and the scattering profile of rather small molecules with its advantage of the wide range of the magnitude of the scattering vector [11]. The latter source is used in the same way as small-angle X-ray scattering, but the advantage is the feasibility of the labeling of deuterated polymer molecules in a soup of other molecules [12].

Chromatography techniques are indispensable for polymer analysis because they separate molecules by molecular weight, monomers with different chemical structures, topological structures, and so on. Size-exclusion chromatography (SEC) or gel-permeation chromatography (GPC) has been widely used to determine molecular weight and molecular weight distribution of polymers [13]. The disadvantage of this method is that the resulting molecular weight depends on the calibration curve, the relation between the molecular weight and the elution volume, which varies with the polymer species of the molecular weight standard used. The situation has changed with the development of online light-scattering detectors, which indicate the molecular weight value at each elution volume. Initially, the instruments had one detector fixed at a low scattering angle (low-angle light scattering = LALS) but instruments with many detectors at different scattering angles (multi-angle light scattering = MALS) became available [14]. Since the MALS detector can determine the radius of gyration from the angular dependence of the scattering intensity at any elution volume, the double-logarithmic plot of the radius of gyration against the molecular weight, sometimes referred to as a "conformation plot", can be easily obtained with a single injection of the sample solution. Online viscosity detectors also became available, allowing the determination of the molecular weight dependence of the intrinsic viscosity with a single injection when used with a LALS or a MALS detector [15].

Separation techniques have also been improved. Supercritical liquid chromatography has improved the molecular-weight resolution of low-molecular weight polymer samples, allowing one peak in a GPC measurement to be split into peaks corresponding to different degrees of polymerization [16]. The development of the chromatography at the critical condition, where the size-exclusion and adsorption effects are balanced, has made it possible to separate polymers by factors other than molecular weight, e.g., topological structures, chemical structures, etc. [17]. The flow-field fractionation has enabled us to separate polymers without using columns [18]. The latter half of Chap. 4 describes the separation of various polymers by chromatography techniques.

Nuclear magnetic resonance (NMR) is currently the most common technique to determine the chemical structure of compounds. This method is applied to determine the chemical structures and the stereoregularity or the copolymer sequence of polymers. The details are shown in the former half of Chap. 4. The other spectroscopic methods, such as infrared and Raman spectroscopies, played important roles to clarify structures of polymers, although they are not contained in this book because they are very specific to the polymer kind.

1.2 What is Characterization?

Generally, polymer analysis refers to determination of the chemical structures of polymers. However, polymers have many types of structures that cannot be expressed by chemical formulas, e.g., molecular weight and its distribution, stereoregularity, monomer sequence, branched structure, chain stiffness, helicity, and so on. Various techniques are used to determine these structures. Such studies are called "characterization". In this book, most of the techniques for characterizing polymers in solution are described.

Characterization techniques for polymeric solid or melt also provide a lot of information about polymer structures. For example, X-ray crystallography is an important technique for determining the crystal structure of polymers. Thermal measurements tell structural changes and the heat incomings and outgoings. Viscoelastic measurements provide information about molecular motion, and so on. However, these are beyond the scope of this book.

There are several books on "polymer characterization" [16–18]. In a typical book, each chapter is devoted to one measuring method, from theoretical background to experimental data treatment. However, for experimentalists, it is difficult to extract the way to reach the required information from such a book. Therefore, the contents of this book are categorized by the information obtained from the experiments. It may help experimentalists quickly find out what they need.

This book consists of four chapters including this Introduction. Chapter 2 describes how to determine molecular weights and their distributions, which was written by Y. N. Chap. 3, written by K. T. and Y. N., is about the molecular size and shape determined by scattering techniques and hydrodynamic measurements. Chapter 4 is about the microstructures determined by NMR and chromatographic methods, written by T. H and A. T., respectively.

References

1. Staudinger H (1920) Über Polymerisation. Ber Dtsch Chem Ges 53:1073–1085. https://doi.org/10.1002/cber.19200530627
2. Debye P (1944) Light Scattering in Solutions. J Appl Phys 15:338–342. https://doi.org/10.1063/1.1707436
3. Zimm BH (1948) The Scattering of Light and the Radial Distribution Function of High Polymer Solution. J Chem Phys 16:1093–1099. https://doi.org/10.1063/1.1746738
4. Svedberg T, Rinde H (1924) The Ultra-Centrifuge, A New Instrument for the Determination of Size and Distribution of Size of Particle in A Microscopic Colloids. J Am Chem Soc 46:2677–2693. https://doi.org/10.1021/ja01677a011
5. Flory PJ (1953) Principles of Polymer Chemistry, Cornell Univ Press, Ithaca, N.Y.
6. Tanford C (1961) Physical Chemistry of Macromolecules, John Wiley & Sons, New York.
7. Norisuye T (1993) Semiflexible polymers in dilute solution. Prog Polym Sci 18:543–584. https://doi.org/10.1016/0079-6700(93)90017-7
8. Yamakawa H, Yoshizaki, T (2016) Helical Wormlike Chains in Polymer Solutions, Springer, Berlin. https://doi.org/10.1007/978-3-662-48716-7

9. Zimm BH, Stockmayer WH (1949) The Dimensions of Chain Molecules Containing Branches and Rings. J Chem Phys 17:1301–1314. https://doi.org/10.1063/1.1747157
10. Berne BJ, Pecora R (1976) Dynamic Light Scattering with Applications to Chemistry, Biology, and Physics, John Wiley & Sons, New York.
11. Glatter O, Kratky O Eds. (1982) Small Angle X-ray Scattering, Academic Press, London.
12. Higgins JS and Benoit HC (1994) Polymers and Neutron Scattering, Oxford Univ. Press, New York.
13. Striegel A, Yau WW, Kirkland JJ, Bly DD (2009) Modern Size-Exclusion Liquid Chromatography: Practice of Gel Permeation and Gel Filtration Chromatography 2nd Ed., John Wiley & Sons, Hoboken, New Jersey. https://doi.org/10.1002/9780470442876
14. Wyatt PJ (1997) Multiangle-Light Scattering: The Basic Tool for Macromolecular Characterization. Instr Sci Tech 25:1–18. https://doi.org/10.1080/10739149709351443
15. Haney MA (1985) The Differential Viscometer. I. A New Approach to the Measurement of Specific Viscosities of Polymer Solutions. J Appl Polym Sci 30:3023–3036. https://doi.org/10.1002/app.1985.070300724
16. Barth HG, Mays JW (Eds) (1991) Modern Methods of Polymer Characterization. John Wiley and Sons, New York.
17. Borsali R, Pecora R (Eds) (2008) Soft-Matter Characterization. Springer, New York. https://doi.org/10.1007/978-1-4020-4465-6
18. Matyjaszewski K, Moller M (Eds) (2012) Polymer Science: A Comprehensive Reference. Elsevier, Amsterdam.

Chapter 2
Molecular Weight and Molecular Weight Distribution

2.1 Overview

Molecular weight is an important factor in controlling the mechanical properties of polymers. It also determines the processability when products are formed. For example, the melt viscosity is known to increase with the molecular weight according to the 3.5th power law. The molecular weight distribution is also important because low molecular weight components act as a plasticizer during polymer processing. For solution processes, the solubility in the solvent changes with the molecular weight. Before discussing measurement methods, this section summarizes basic knowledge about the average molecular weights and the molecular distribution.

2.1.1 Average Molecular Weights

Most of the commonly used polymers are obtained by polymerization of monomers. If the molecule contains n monomers, it is called n-mer. Typical synthetic polymers are composed of various n-mers. In other words, the polymer has a molecular weight distribution. Therefore, when measuring the molecular weight of a polymer compound, the obtained molecular weight is an averaged value. The type of averaged molecular weight obtained depends on the measurement method. Here, we introduce typical average molecular weights calculated by various weight functions.

When the polymer compound contains m_i moles of ith component with molecular weight M_i, the number fraction (or the mole fraction) of the ith component is expressed as

---

Author: Y. Nakamura.

Y. Nakamura et al., *Characterization of Polymers in Solution*,
https://doi.org/10.1007/978-981-96-6677-5_2

$$n_i = \frac{m_i}{\sum_j m_j} \tag{2.1.1}$$

Here, the summation is taken over all the components. The weight fraction w_i is defined by

$$w_i = \frac{m_i M_i}{\sum_j m_j M_j} = \frac{n_i M_i}{\sum_j n_j M_j} \tag{2.1.2}$$

Using these fractions, the following average molecular weights are defined.

a. **Number-Average Molecular Weight**

Using n_i as the weighting function, the number-average molecular weight M_n is defined by

$$M_\mathrm{n} = \sum_i n_i M_i \tag{2.1.3}$$

M_n can be measured by colligative properties such as osmotic pressure. The above equation is also expressed using w_i as follows:

$$M_\mathrm{n} = \left(\sum_i \frac{w_i}{M_i} \right)^{-1} \tag{2.1.4}$$

b. **Weight-Average Molecular Weight**

The molecular weight averaged by the weight fraction is called as weight-average molecular weight, which is defined by

$$M_\mathrm{w} = \frac{\sum_i m_i M_i^2}{\sum_j m_j M_j} = \frac{\sum_i n_i M_i^2}{\sum_j n_j M_j} = \sum_i w_i M_i \tag{2.1.5}$$

The typical method to obtain M_w is light scattering measurement.

c. **z-Average Molecular Weight**

The z-average molecular weight M_z is defined by

$$M_\mathrm{z} = \frac{\sum_i m_i M_i^3}{\sum_j m_j M_j^2} = \frac{\sum_i n_i M_i^3}{\sum_j n_j M_j^2} = \frac{\sum_i w_i M_i^2}{\sum_j w_j M_j} \tag{2.1.6}$$

M_z can be measured by ultracentrifugation.[1]

---

[1] The subscript "z" comes from the German word for centrifugation, *zentrifugation.*

2.1.2 Polydispersity Index

The ratio M_w/M_n is generally used as the polydispersity index. The reason why it indicates the molecular weight dispersity can be understood by calculating $M_w/M_n - 1$:

$$\frac{M_w}{M_n} - 1 = \frac{\sum_i n_i M_i^2 - \left(\sum_i n_i M_i\right)^2}{\left(\sum_j n_j M_j\right)^2} = \frac{\overline{M^2} - \overline{M}^2}{\overline{M}^2} \tag{2.1.7}$$

The overlines indicate the number averaged value. The above equation shows that the quantity is equal to the variance used in the statistics divided by the square of the (number average) mean value. When the compound consists of monodisperse molecules, the dispersity is zero, therefore $M_w/M_n = 1$. The wider the molecular weight dispersion is, the larger the M_w/M_n value.

2.2 Molecular Weight

2.2.1 Osmotic Pressure

Osmotic pressure measurement is the oldest technique for molecular weight measurements of large molecules. The principle is simple. As shown in Fig. 2.1, the osmotic pressure cell is separated into two parts by a semipermeable membrane, through which the solvent molecules can pass but the solute molecules cannot. One side of the cell is filled with the solvent and the other with the solution. Due to the dilution effect, the solvent penetrates into the solution side until the equilibrium is attained. In equilibrium, there is a pressure difference between the solution and the solvent. This is osmotic pressure. Primarily, the osmotic pressure Π is expressed by the following van't Hoff equation

$$\frac{\Pi}{RTc} = \frac{1}{M} \tag{2.2.1}$$

where R is the gas constant, T is the absolute temperature, c is the mass concentration of the solute, and M is the solute molecular weight. The above equation is available only in the limit of $c \to 0$. At finite concentrations, the concentration dependence can be expressed by the power series of c as

$$\frac{\Pi}{RTc} = \frac{1}{M} + A_2 c + A_3 c^2 + \cdots \tag{2.2.2}$$

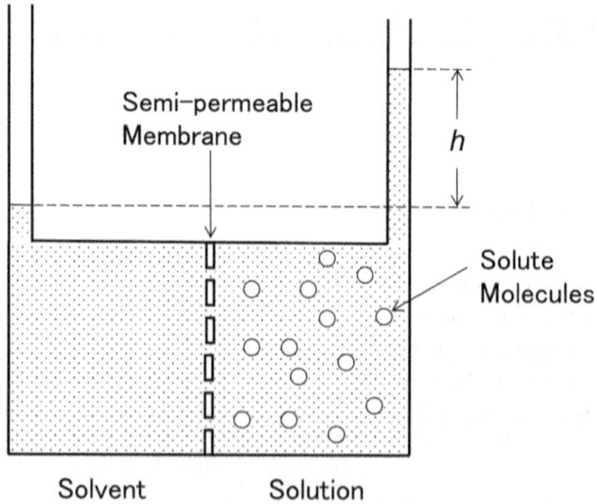

Fig. 2.1 Schematic representation of osmotic pressure measurements

when c is small. This equation is called the virial expansion, where A_2 and A_3 are called as the second and the third virial coefficients, respectively. The former coefficient is used for the index of the goodness of the solvent. If A_2 is positive, the solvent is good. Conversely, if A_2 is negative, the solvent is poor. When A_2 vanishes, the solvent is called the theta solvent. The solution of $A_2 = 0$ is in the theta condition, where the intra- and intermolecular interactions apparently vanish.

When the solute is polydisperse in molecular weight, $M_{\rm n}$ is obtained instead of M, because of the following reason. If the mass concentration of the ith solute component with the molecular weight M_i is c_i, the contribution of this component to Π is written as

$$\Pi_i = \frac{RTc_i}{M_i} \tag{2.2.3}$$

The total osmotic pressure may be written as the sum of Π_i, then

$$\Pi = \sum_i \Pi_i = \sum_i \frac{RTc_i}{M_i} \tag{2.2.4}$$

Since $c_i/c = w_i$, we obtain

$$\frac{\Pi}{RTc} = \sum_i \frac{w_i}{M_i} = \frac{1}{M_{\rm n}} \tag{2.2.5}$$

When the osmotic pressure is measured by an apparatus as shown in Fig. 2.1, the measurement takes several hours (or days) to reach equilibrium. To obtain equilibrium more quickly, the currently available osmotic pressure apparatuses have the structure as shown in Fig. 2.2. The cell is separated by a semipermeable membrane

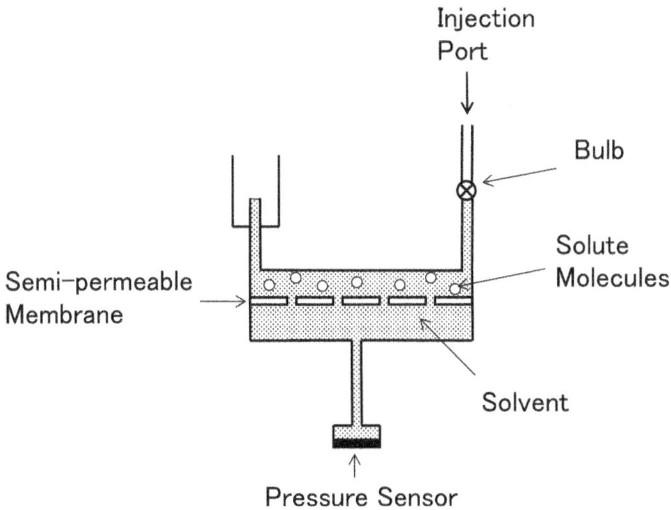

Fig. 2.2 Schematic representation of the commercially available osmotic pressure apparatus

into the upper and the lower parts and the pressure of the lower part filled with the solvent is monitored by a pressure sensor. At first, the upper part is filled with the solvent, the indicated pressure is adjusted to zero, and then the upper part is substituted by the solution. Because the pressure of the upper part is atmospheric, the pressure of the lower part is reduced by the osmotic pressure. The absolute value of the reduced pressure after reaching the equilibrium is equated to Π. The time needed to attain the equilibrium is typically 10 min for this type of apparatus.

Figure 2.3 shows an examples of osmotic pressure data for polyisobutylene (PIB) [1], where Π/RTc for PIB in cyclohexane increases with increasing c showing that A_2 is positive, indicating that the solvent is good. For PIB in benzene, $A_2 = 0$ and the solution is in the theta state.

In Fig. 2.3a, the data for PIB in cyclohexane cannot be fitted by a straight line. In such cases, the square root plot is used. By taking square root on both sides of Eq. 2.2.2, the osmotic pressure is expressed as follows:

$$\left(\frac{\Pi}{RTc}\right)^{1/2} = \frac{1}{M^{1/2}} + \frac{1}{2}A_2M^{1/2}c + \frac{1}{2}A_2^2M^{3/2}\left(g - \frac{1}{4}\right)c^2 + \cdots \tag{2.2.6}$$

Here, g is the reduced third virial coefficient defined by

$$g \equiv \frac{A_3}{A_2^2M} \tag{2.2.7}$$

It is known that g for typical flexible polymers in good solvents is around 1/4 [2]. Therefore, the square-root plot gives better linearity, as shown in Fig. 2.3b.

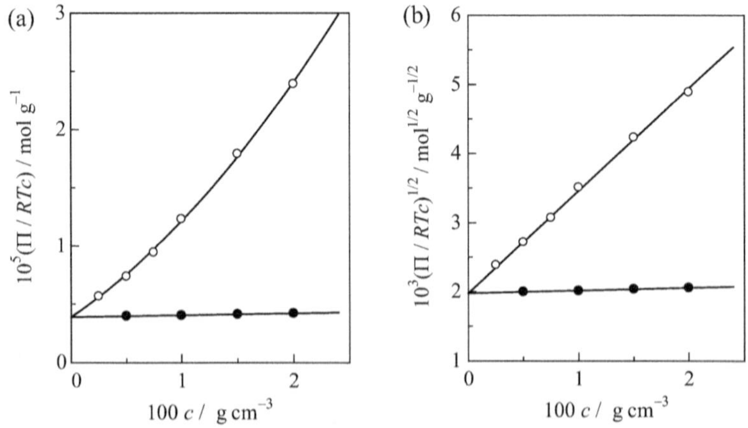

Fig. 2.3 Examples of osmotic pressure data for polyisobutylene in cyclohexane [1] (unfilled circles) and benzene (filled circles)

The molecular weight range being measurable by the osmotic pressure ranges from several thousands to about a million. Currently, molecular weight measurements have mostly shifted to light scattering measurements. However, osmotic pressure measurements still have advantages to measure molecular weights for polymers absorbing visible light.

2.2.2 *Vapor Pressure Osmometry*

Vapor pressure osmometry (VPO) is not based on the osmosis but on the vapor pressure change. When the ideal solution consisting of a solvent (component 1) and a nonvolatile solute (component 2) is in equilibrium with the gas phase consisting of the solvent vapor, the pressure of the gas phase p is related with the vapor pressure of the pure solvent $p_1{}^\circ$ by Raoult's rule

$$p = p_1^\circ x_1 \tag{2.2.8}$$

where x_1 is the mole fraction of component 1 in the solution. Since the mole fraction of the solute x_2 $(= 1 - x_1)$ can be approximately depicted as V_1c/M with the molar volume of the solvent V_1, the mass concentration of the solute c, and the molecular weight M, the difference Δp between $p_1{}^\circ$ and p is expressed as

$$\frac{\Delta p}{p_1^\circ V_1 c} = \frac{1}{M} \tag{2.2.9}$$

although it is difficult to measure Δp precisely because it is small when M is large.

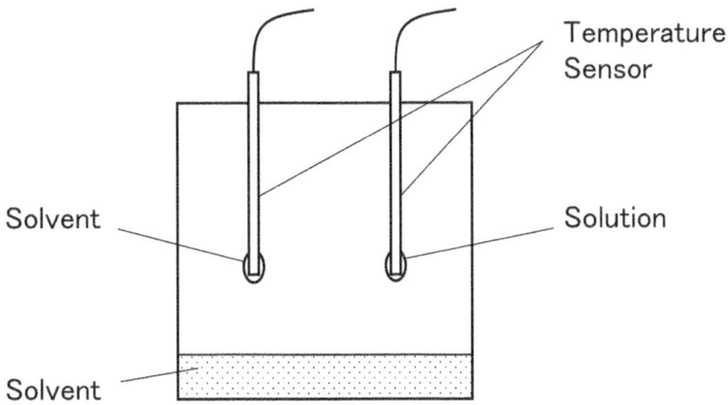

Fig. 2.4 Schematic representation of vapor pressure osmometry

The schematic representation of the VPO measurement is shown in Fig. 2.4, where two temperature sensors are placed in a chamber filled with the solvent vapor. Each edge of the sensors is wetted with the solution and the solvent droplets, respectively, and the temperature difference ΔT of them is measured. Since ΔT is proportional to Δp, the above equation can be rewritten as

$$\frac{K_v \Delta T}{c} = \frac{1}{M} \tag{2.2.10}$$

where K_v is a constant that contains the proportional constant between ΔT and Δp with $p_1{}^\circ V_1$ in Eq. 2.2.9. Using a standard substance with known M, K_v is determined in advance. Because the right-hand side of the above equation is written as a power series of c as is the case for the osmotic pressure, $K_v \Delta T / c$ is plotted against c and the $1/M$ is obtained as the value extrapolated to $c = 0$. For solutes having molecular weight distribution, the obtained molecular weight is M_n. The range of the molecular weight determined by VPO is from ca. 50 to tens of thousands.

2.2.3 Sedimentation Equilibrium

When a solution containing a solute having higher density than the solvent is in a centrifugal field, the solute molecules move away from the center of rotation by the centrifugal force. For analytical ultracentrifugation, special instruments are used. The rotor for the analytical ultracentrifugation has holes for inserting centerpieces with a sector to contain the solution as shown in Fig. 2.5a. The upper and lower sides of the centerpiece are capped with transparent windows which enable the light to pass through the solution. By using the ultraviolet absorption or the interference fringe (using the centerpiece having two sectors for the solution and the solvent, respectively), the solute concentration $c(r)$ is monitored as a function of the distance

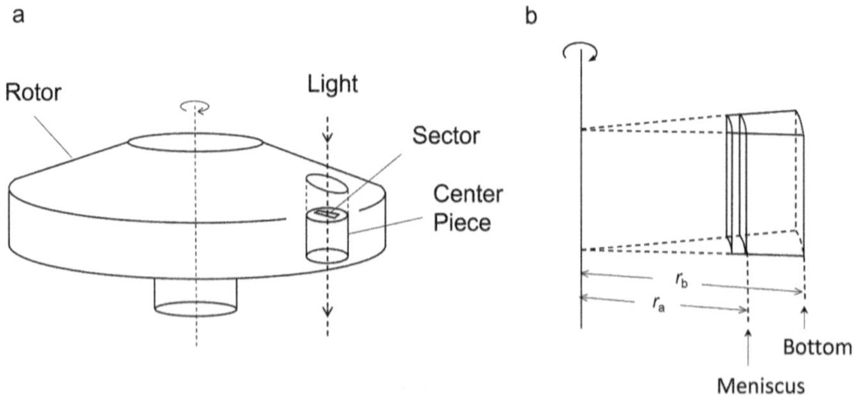

Fig. 2.5 Schematic representation of the analytical ultracentrifugation (**a**) and the sector for the solution (**b**)

r from the center of rotation. Although the original centrifugal instrument can rotate only one solution per measurement, modern synchronization technology allows to rotate up to eight solutions and to obtain the concentration profile of each solution at a time.

In the sedimentation equilibrium method, the centrifuge is operated at a relatively low angular velocity ω, where the sedimentation and diffusion of the solute molecules are balanced. A typical example of sedimentation equilibrium measurement is shown in Fig. 2.6, where $A(r)$ is the UV absorbance proportional to $c(r)$. At the initial stage, the solution concentration is independent of r as shown in Fig. 2.6a. At this condition, the proportional constant between c and the absorbance is determined using the initial solute concentration c_0. After increasing the rotation velocity, the solution concentration gradually increases with r and reaches equilibrium as shown in Fig. 2.6b, where ln $A(r)$ is plotted against r^2. When the solute is monodisperse in molecular weight, the graph is represented by a straight line. However, when the solute is polydisperse in molecular weight, the graph is curved as shown in the figure. The position of the meniscus and the bottom of the solution is represented by r_a and r_b, respectively (see Fig. 2.5b). Fitting the data by an appropriate curve and reading the values at r_a and r_b, the concentrations at the meniscus and the bottom, $c(r_a)$ and $c(r_b)$ ($A(r_a)$ and $A(r_b)$ in Fig. 2.6), respectively, are determined. From these values, the apparent molecular weight M_{app} is calculated by the following equation.

$$M_{app}^{-1} = \frac{\lambda c_0}{c(r_b) - c(r_a)} \tag{2.2.11}$$

where c_0 is the initial solution concentration before centrifugation. In the above equation, λ is a constant defined by

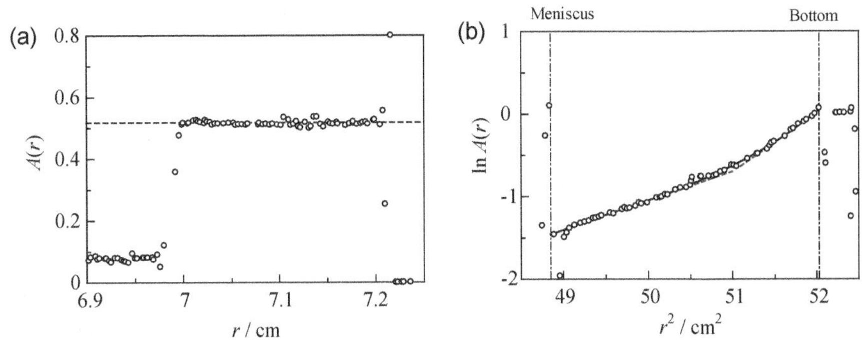

Fig. 2.6 Example of the UV absorbance $A(r)$ as a function of the distance r from the center of rotation at low angular velocity (3000 r.p.m.) (**a**) and at high angular velocity (17,000 r.p.m. after 23 h) (**b**) for a colored polymer in H_2O [2]

$$\lambda = \frac{\omega^2(1-\overline{v}\rho_0)(r_b^2 - r_a^2)}{2RT} \tag{2.2.12}$$

Here, ρ_0 is the solvent density, $\overline{v}$ is the partial specific volume of the solute, R is the gas constant, and T is the absolute temperature.

M_{app} is expressed by a power series of the average concentration of $c(r_a)$ and $c(r_b)$, i.e., $\overline{c} = [c(r_a) - c(r_b)]/2$, as follows

$$M_{app}^{-1} = M_w^{-1} + 2A_2\overline{c} + \cdots \tag{2.2.13}$$

In an actual molecular weight determination, several solutions with different c_0 are prepared and $M_{app}{}^{-1}$ is obtained as a function of $\overline{c}$. Fitting the plot of $M_{app}{}^{-1}$ against $\overline{c}$ by a straight line, the slope and the intercept at $\overline{c} = 0$ are equated to $2A_2$ and $M_w{}^{-1}$, respectively.

To obtain M_z, the following value is calculated from the slope at r_a and r_b, i.e., $(d \ln c/dr^2)_a$ and $(d \ln c/dr^2)_b$, respectively, in the plot of $\ln c(r)$ versus r^2 (dashed lines in Fig. 2.5b):

$$\begin{aligned} Q &\equiv \frac{(c_b - c_a)^2}{c_0(r_b^2 - r_a^2)[c_b(d \ln c/dr^2)_b - c_a(d \ln c/dr^2)_a]} \\ &= \frac{M_z}{M_w}(1 + 2M_wA_2\overline{c} + \cdots) \end{aligned} \tag{2.2.14}$$

Plotting Q against $\overline{c}$ and extrapolating to $\overline{c} = 0$, M_z/M_w is obtained.

2.2.4 Light Scattering

Light scattering measurement is one of the most reliable methods to determine molecular weight of polymers. A schematic representation is given in Fig. 2.7. A monochromatic light beam, such as laser, is inserted into a cylindrical glass cell from its side, so that the beam passes through the center of the cell. The intensity of the scattered light from a volume (called the scattering volume) of the irradiated part of the solution is measured by a detector, usually a photomultiplier or a photodiode. The angle between the direction of the incident beam and the direction of the scattering light, i.e., the line connecting the center of the cell and the detector, is called the scattering angle θ. Such a type of instrument as shown in the figure, of which the detector is mounted on an arm and can be rotated around the cell to change the scattering angle, is called goniometers. From the scattering intensity $I(\theta)$ at θ from the scattering volume V, the Rayleigh ratio R_θ is calculated according to the following equation

$$R_\theta = \frac{r^2 I(\theta)}{I^\circ V(1+\cos^2\theta)} \tag{2.2.15}$$

when an unpolarized incident light is used. Here, r is the distance from the scattering volume to the detector and I° is the intensity of the incident light. Currently, vertically polarized light is used for typical measurements. In such cases, R_θ is given by

$$R_\theta = \frac{r^2 I(\theta)}{I^\circ V} \tag{2.2.16}$$

Usually, light scattering data for polymer solutions are described by the reciprocal intensity, Kc/R_θ, where c is the solute mass concentration and K is the scattering constant defined by

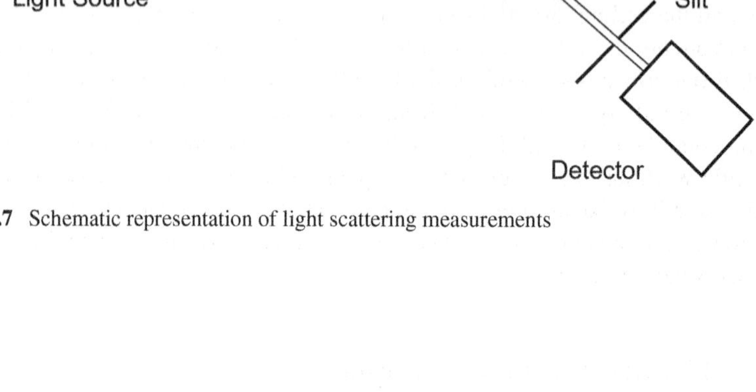

Fig. 2.7 Schematic representation of light scattering measurements

$$K = \frac{4\pi^2 n_0^2}{N_A \lambda_0^4}\left(\frac{\partial n}{\partial c}\right)^2 \tag{2.2.17}$$

for vertically polarized light. Here, N_A is the Avogadro constant, λ_0 is the wavelength of the incident light in the vacuum, and $(\partial n/\partial c)$ is the specific refractive index increment of the solution, which is the slope of the solution refractive index n plotted against c.

Literature values of $(\partial n/\partial c)$ for various polymer-solvent systems are summarized in *Polymer Handbook* [3]. If the $(\partial n/\partial c)$ value for the studying polymer-solvent system and at the wavelength λ_0 of your light-scattering instrument can be found, there is no need to measure it. If there are several values of $(\partial n/\partial c)$ for the current polymer-solvent system but not for the aimed λ_0, the needed $(\partial n/\partial c)$ value can be obtained by interpolation or extrapolation assuming the linear relation between $(\partial n/\partial c)$ and $1/\lambda_0{}^2$. However, if no literature values can be found for the polymer-solvent system under study, it must be measured in the following manner.

Preparing from four to seven solutions with different concentrations, the difference of the refractive index, Δn, between each solution and the solvent is measured by a differential refractometer or online refractive index detector with the same λ_0 as that for the light scattering instrument. Figure 2.8a shows an example of Δn for poly(ethylene oxide) solutions in water injected to the online refractometer changing in the order of the concentration. The height of the plateau for each concentration is converted to Δn by multiplying the instrumental constant, which is determined in advance using standard solutions of known $(\partial n/\partial c)$. Figure 2.8b shows the resulted plot of Δn against c. The $(\partial n/\partial c)$ value is obtained from the slope of the straight line fitted to the plot.

Light scattering values (Kc/R_θ) extrapolated to $\theta = 0$ is written as

$$\left(\frac{Kc}{R_\theta}\right)_{\theta=0} = \frac{1}{M_w} + 2A_2c + 3A_3c^2 + \cdots \tag{2.2.18}$$

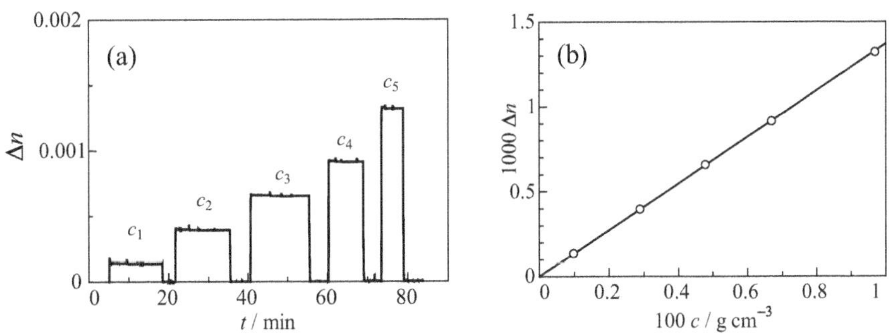

Fig. 2.8 Example of $(\partial n/\partial c)$ data for poly(ethylene oxide) in water (unpublished data)

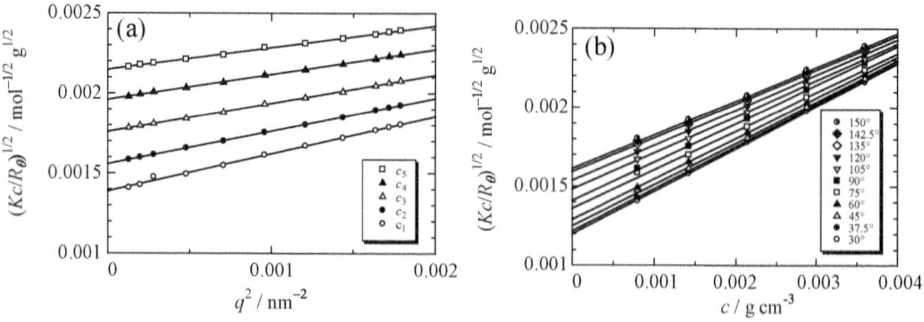

Fig. 2.9 Examples of extrapolation of $(Kc/R_\theta)^{1/2}$ to $q^2 = 0$ and $c = 0$ for polystyrene in benzene (unpublished data)

which is similar to Eq. 2.2.2.[2] On the other hand, (Kc/R_θ) extrapolated to $c = 0$ is expressed as

$$\left(\frac{Kc}{R_\theta}\right)_{c=0} = \frac{1}{M_\mathrm{w}}\left(1 + \frac{\langle S^2\rangle}{3}q^2 + \cdots\right) \tag{2.2.19}$$

Here, $\langle S^2\rangle$ is the mean-square radius of gyration and q is the magnitude of scattering vector expressed as

$$q = \frac{4\pi n_0}{\lambda_0}\sin\frac{\theta}{2} \tag{2.2.20}$$

In Fig. 2.9, an example of extrapolations of $(Kc/R_\theta)^{1/2}$ to $q^2 = 0$ and $c = 0$ are shown. Using the square root plots, the linearity of the data is much improved when the solute is chain molecules. For these plots, the equations are changed as follows:

$$\left(\frac{Kc}{R_\theta}\right)^{1/2}_{\theta=0} = \frac{1}{M_\mathrm{w}^{1/2}}\left[1 + A_2M_\mathrm{w}c + \frac{3}{2}A_2^2M_\mathrm{w}^2\left(g - \frac{1}{3}\right)c^2 + \cdots\right] \tag{2.2.21}$$

$$\left(\frac{Kc}{R_\theta}\right)^{1/2}_{c=0} = \frac{1}{M_\mathrm{w}^{1/2}}\left(1 + \frac{\langle S^2\rangle}{6}q^2 + \cdots\right) \tag{2.2.22}$$

Comparing Eq. 2.2.21 with Eq. 2.2.6, the latter equation shows that the concentration dependence of osmotic pressure data gives better linearity when $g = 1/4$, while the former equation shows that of light scattering data has better linearity when $g = 1/3$.

[2] For polydisperse solutes, the virial coefficients determined by light scattering are not exactly equal to those for osmotic pressure. See, e.g., *Thermodynamics of Polymer Solutions*, Michio Kurata (Translated from the Japanese by Hiroshi Fujita), MMI Press, Chur, 1982.

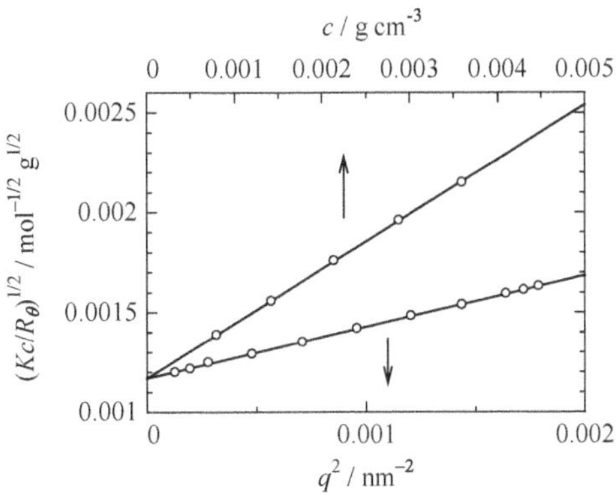

Fig. 2.10 Plots of $(Kc/R_\theta)_{\theta=0}{}^{1/2}$ against c (upper) and $(Kc/R_\theta)_{c=0}{}^{1/2}$ against q^2 (lower)

Plots of $(Kc/R_\theta)_{\theta=0}{}^{1/2}$ against c and $(Kc/R_\theta)_{c=0}{}^{1/2}$ against q^2 are shown in Fig. 2.10, which give A_2 and $\langle S^2\rangle$ from the slopes, respectively, and M_w from the intercept. More details of light scattering measurements are given in Sect. 3.2.1.

2.2.5 *Intrinsic Viscosity*

The viscosity of dilute polymer solutions can be measured by the Ubbelohde capillary viscometer, which is schematically shown in Fig. 2.11. The sample solution is inserted from C into the vessel E. Sealing A, the solution is sucked from B until the liquid level passes over the line L1. Next, opening A, the solution below the lower end of the capillary is dropped. This ensures that the lower liquid level of the measured solution is always kept at this height. By opening B, the solution starts to fall freely. The flow time t, which is the time that the solution liquid level passes from L1 to L2, is measured. To control the temperature, the viscometer is usually dipped in a water bath.

Since the liquid viscosity is proportional to the flow time, the relative solution viscosity to that of the solvent is calculated by the following equation.[3]

$$\eta_r = t/t_0 \tag{2.2.23}$$

for dilute solutions, where t_0 is the flow time of the solvent. The specific viscosity is defined as follows:

$$\eta_{sp} = (t - t_0)/t_0 = \eta_r - 1 \tag{2.2.24}$$

The intrinsic viscosity $[\eta]$ is determined by extrapolating η_{sp}/c, $(\ln \eta_r)/c$, or $[2(\eta_{sp} - \ln \eta_r)]^{1/2}/c$ to $c = 0$ according to the following equation valid for small c.

[3] Exactly η_r is equal to $(\rho t)/(\rho_0 t_0)$ where ρ and ρ_0 mean the densities of the solution and solvent, respectively, while ρ/ρ_0 can be approximated to 1 for dilute solutions.

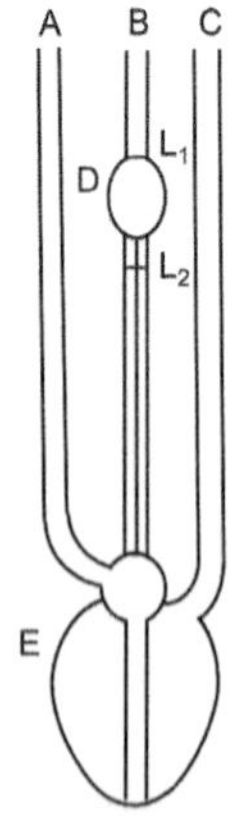

Fig. 2.11 Schematic representation of the Ubbelohde viscometer

$$\frac{\eta_{sp}}{c} = [\eta] + k'[\eta]^2 c + \cdots \tag{2.2.25}$$

$$\frac{\ln \eta_r}{c} = [\eta] - \left(\frac{1}{2} - k'\right)[\eta]^2 c + \cdots \tag{2.2.26}$$

The following equation is also valid to determine $[\eta]$.

$$\frac{[2(\eta_{sp} - \ln \eta_r)]^{1/2}}{c} = [\eta] + \left(k' - \frac{1}{3}\right)[\eta]^2 c + \cdots \tag{2.2.27}$$

Here, k' is called the Huggins constant. The concentration dependence of η_{sp}/c, $(\ln \eta_r)/c$, and $[2(\eta_{sp} - \ln \eta_r)]^{1/2}/c$ for a polystyrene sample in benzene (good solvent) and cyclohexane (Θ solvent) [4] are shown in Fig. 2.12. The curvature of the fitted lines for the cyclohexane solutions indicates that the effect of the c^2 term cannot be ignored. For flexible polymers, k' is about 0.3 for good solvent systems and is about 0.5 for theta solvent systems.

Molecular weight dependence of $[\eta]$ of polystyrene in benzene and cyclohexane [4] is shown in Fig. 2.13. For $M_w \geq 10^4$, the plot for each solvent system can be fitted by a straight line, indicating that $[\eta]$ is described as

$$[\eta] = KM^{\alpha} \tag{2.2.28}$$

This relation is called Mark-Houwink-Sakurada equation. For a lot of polymer-solvent systems, K and α values are known [3]. If the K and α for the polymer solvent system are known, the molecular weight of an unknown sample can be determined by measuring $[\eta]$ using the above equation. The molecular weight obtained from $[\eta]$ is called the viscosity average molecular weight, which is given by

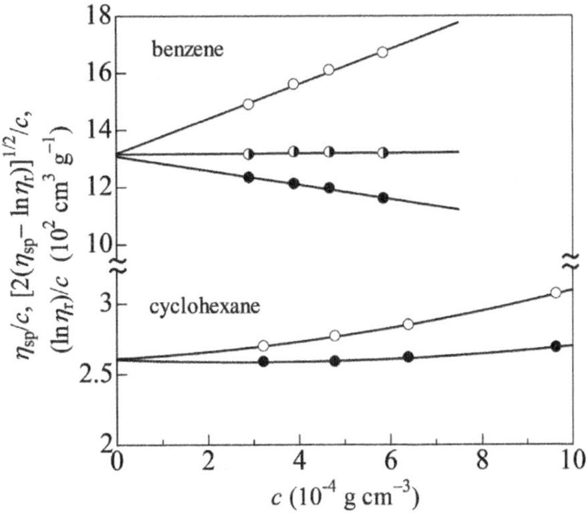

Fig. 2.12 Examples of viscosity data for polystyrene in the solvents indicated [4]. Unfilled, filled, and half-filled circles show η_{sp}/c, $(\ln \eta_r)/c$, and $[2(\eta_{sp} - \ln \eta_r)]^{1/2}/c$, respectively

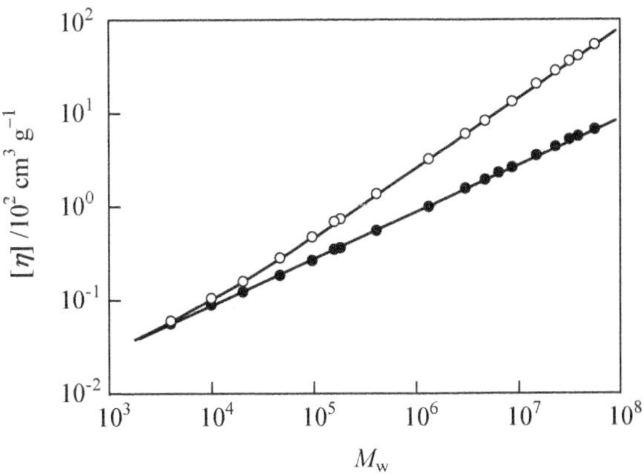

Fig. 2.13 Intrinsic viscosities for polystyrene in benzene at 25.0 °C (unfilled circles) and cyclohexane at 34.5 °C (filled circles) [4] double-logarithmically plotted against M_w

$$M_v = \left(\sum_i w_i M_i^{\alpha} \right)^{1/\alpha} \tag{2.2.29}$$

When $\alpha = 1$, M_v becomes identical to M_w.

Flory and Fox [5] proposed the following relation between $[\eta]$ and $\langle S^2 \rangle$.

$$[\eta] = \Phi \frac{(6\langle S^2 \rangle)^{3/2}}{M} \tag{2.2.30}$$

Here, Φ is called the Flory-Fox factor. For flexible polymers in theta solvents, Φ is about 2.5×10^{23} mol^{-1}, while Φ is about 2.0×10^{23} mol^{-1} for flexible polymers in good solvents (See Fig. 3.16).

2.3 Molecular Weight Distribution

2.3.1 *Size-Exclusion Chromatography*

2.3.1.1 Calibration Curve Method

When a polymer solution flows into a column packed with gel beads having small cavities on the surface, small polymer molecules are trapped into the cavities, while large ones are excluded. By this size-exclusion effect, larger molecules flow out faster. Chromatography utilizing this effect is called size-exclusion chromatography (SEC) or gel-permeation chromatography (GPC).

The SEC chromatograms for poly(hexyl isocyanate) (PHIC) [6] in tetrahydrofuran (THF) are shown in Fig. 2.14. The upper part is the plot of log M against the elution volume V_e at the peak top, where V_e equals the elution time multiplied by the flow rate. The relation, called the calibration curve, between log M and V_e may be fitted by a straight line or a polynomial curve.

Usually, refractive index, ultraviolet absorption, or evaporative light-scattering detectors are used for SEC. In the chromatogram obtained by these detectors, the peak height is proportional to the mass concentration of the solute. Therefore, at a certain V_e of the peak, the weight fraction, w_i, of the component (denoted as the ith component) can be obtained by dividing the signal height (multiplied by the interval of V_e) by the integrated signals, while the molecular weight M_i for this component can be evaluated from V_e using the calibration curve. Using these values, M_n, M_w, and M_z can be calculated by Eqs. 2.1.4–2.1.6, respectively. As is mentioned in Sect. 2.1.2, M_w/M_n is the measure of molecular weight distribution. The distribution of molecular components is also visualized by the peak shape.

To obtain the calibration curve, standard samples, typically polystyrene, with known M are usually used. However, the molecular weight values calculated by using the calibration curve determined based on standard polystyrene are precise only for

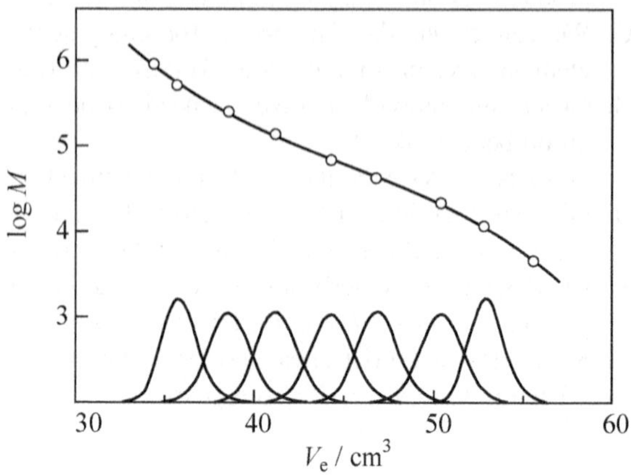

Fig. 2.14 Calibration curve (upper) and SEC chromatograms (lower) for PHIC in THF [6] (Chromatograms for the lowest and highest molecular weight samples are omitted.)

polystyrene samples. Molecular weight values for other samples determined using the calibration curve based on polystyrene standards are merely relative values and not exact.

2.3.1.2 Universal Calibration Curve Method

The method using the universal calibration curve compensates for the defect of the molecular weight determination using the calibration curve based on polystyrene standards. By the following equation based on the Einstein equation of $[\eta]$ for spherical molecules, the hydrodynamic volume V_H of a molecule is defined.

$$[\eta] = \frac{5}{2}\frac{N_A}{M}V_H \tag{2.3.1}$$

The above equation tells that $[\eta]M$ reflects the volume of the polymer molecule. Since the ability of polymer molecules to enter the holes on the gel beads depends on the size of the molecule, the elution volume is expected to be a function of $[\eta]M$. Figure 2.15 shows the plot of log $[\eta]M$ vs V_e for various kinds of polymers [7]. In spite of the variety of chemical structures (PS, polystyrene; PMMA, PVC, poly(vinyl chloride), PB, polybutadiene, PPS, poly(phenyl siloxane)), branched structures (comb, star), and copolymer structures (graft, random), all the data points are described by a single curve. This curve is called the universal calibration curve. The universal curve can be obtained by determining V_e for standard samples of known M and the M value multiplied by the $[\eta]$ calculated by the MHS relation.

When the MHS relation is not known for the polymer solved in the eluent solvent, an online viscometer is available. The structure of an online viscometer is schematically shown in Fig. 2.16 [8]. The flow from SEC is inserted into the bridge circuit, which is constructed by registers (from R_1 to R_4) made of coiled pipes. At the entrance of the circuit, the flow is divided into two ways. One flow goes straightly into the registers and the other goes into the same registers with a column to delay the flow. Using differential pressure gauges, the pressure difference between the entrance and the exit of the circuit, p_I, and that between the middle of the separated two flows, p_D (see the figure), are measured. From these pressure values, η_{sp} is calculated by the following equation [8].

$$\eta_{sp} = \frac{4p_D}{p_I - 2p_D} \tag{2.3.2}$$

Using a concentration detector, RI or UV, serially connected to the viscometer, η_{sp}/c is obtained. Ignoring the concentration effect, η_{sp}/c, or more preferably $[2(\eta_{sp} - \ln \eta_r)]^{1/2}/c$, is regarded as $[\eta]$.

The M_n and M_w values obtained from the universal calibration curve are more accurate than the values based on the calibration curve determined by polymer standards different from the test sample.

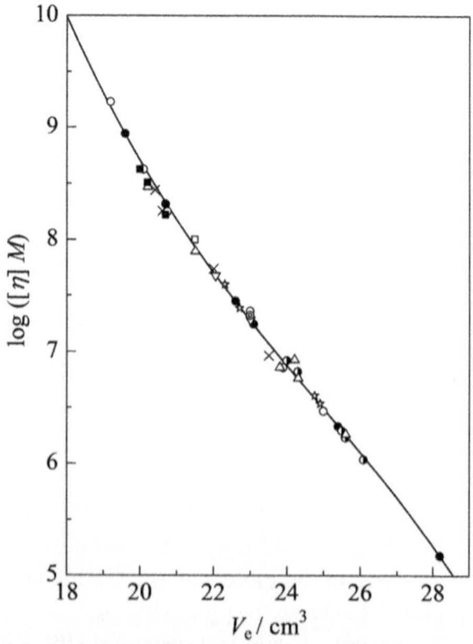

Fig. 2.15 Example of the universal calibration curve [7], unfilled circles, PS; cross, PMMA; half-filled circles, PVC; squares, PB; filled squares, PPS; diamonds, comb-PS; stars, star-PS; triangles down, PMMA-*g*-PS; triangles up, (PMMA-*g*-PS)-*b*-PS-*b*-(PMMA-*g*-PS); double circles, PS-*r*-PMMA

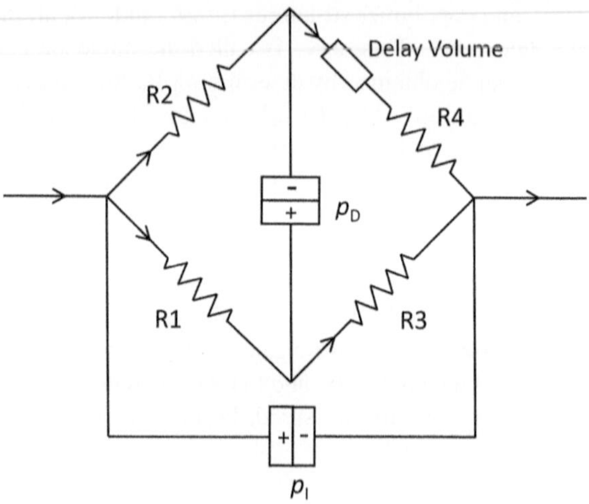

Fig. 2.16 Schematic representation of the differential pressure viscometer [8]

2.3.1.3 SEC with Online Light Scattering Detector

Connecting a low-angle light scattering (LALS) or a multi-angle light scattering (MALS) detector to SEC, the molecular weight (exactly M_w) of the solute polymer at each V_e can be known without using the calibration curve. When a MALS detector is available, the mean-square radius of gyration $\langle S^2 \rangle$ at each V_e also can be measured. An example for this SEC-MALS method is given in Fig. 2.17, in which the peaks

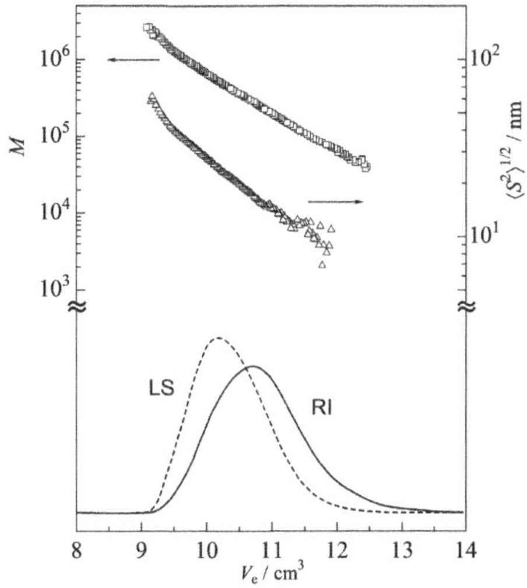

Fig. 2.17 Molecular weights and radii of gyration for a polyethylene sample determined by SEC-MALS-RI system [9]

show the signals from RI and MALS detectors for a polyethylene sample in THF [9]. The upper part of the figure shows the determined values of M_w and $\langle S^2\rangle^{1/2}$ at each V_e. The relation between $\langle S^2\rangle$ and M_w for each sample can be obtained by only single injection of the solution into SEC. How to analyze the $\langle S^2\rangle - M_w$ data will be described in Sect. 3.4.

2.3.2 *Mass Spectroscopy*

Typical mass spectroscopic methods to determine molecular weight of polymers are the electro-spray ionization (ESI) and the matrix-aided laser desorption ionization (MALDI) techniques. In the former method, eluent flowed out from a liquid chromatography is sprayed by a nebulizer. The aerosol is charged and introduced into a chamber to vaporize the solvent. In this process, the polymer molecules are charged and accelerated by the high voltage applied. The final velocity depends on the number of charges on a molecule and the molecular mass. The disadvantage of this method is that it produces complicated mass spectrometry data because of the generation of multivalent molecular ions.

In the MALDI method, a small amount of the solution containing the sample polymer, a matrix compound, and an ionic substance is dropped on a metal sample plate to obtain a thin film. The matrix compounds are low molecular weight substances which can be easily ionized in a high vacuum by irradiating laser beam. The sample plate is placed in a high vacuum chamber, where a high voltage is applied, and irradiated with a laser beam to ionize the sample. Ionized polymer molecules are accelerated toward the electrode of opposite polarity. After passing the electrode, the ionized polymer molecules freely fly over a certain distance and are detected by an analyzer, usually an electro multiplier. Because lighter molecules fly faster and heavier molecules slower, molecules are detected in the order of molecular weight. The molecular weight of an unknown sample is determined from the relationship between the molecular weight and the flight time determined in advance by molecular weight standards. This method is called time-of-fly mass spectroscopy (TOF-MS).

For a typical measurement, matrix solution with the concentration ca. 10 mg/mL, polymer solution with the concentration ca. 1mg/mL, and solution of metal ionic compound, such as silver trifluoroacetic acid, with the concentration 2 mg/mL are mixed with the volume ratio 10:1:1, and then 0.2–1.0 μL of the mixture solution is spotted on the sample plate of the instrument. After the solvent is fully evaporated, the plate is set into the MALDI-TOF-MS instrument. Commonly used matrix compounds are Dithranol, 2,5-dihydroxy benzoic acid (DHB), 2-(4-hydroxy-phenylazo) benzoic acid (HABA), and so on. Usually, the matrix compound must be optimized according to the polymer to be analyzed [10].

An example of MALD-TOF-MS spectrum for a polystyrene sample is shown in Fig. 2.18. The signal from the polymer material is separated into several peaks having different degrees of polymerization. Precisely, each peak in the figure is further separated into multiple peaks because of the included isotopes, H^2, C^{13}, and so on, when magnified. Since the signal of a typical MS spectrum is proportional to the number of molecules, the number fraction n_i of polymer with polymerization degree i is obtained. Therefore, M_n, M_w, and M_z can be calculated from the MS data using Eqs. 2.1.3–2.1.6.

The disadvantage of MALDI-TOF-MS measurement is that the sensitivity is lowered as the molecular weight increases. Figure 2.19 shows the relative integrated intensity $R(M)$ for narrow-dispersed polystyrene samples plotted against the number-average molecular weight. It shows that $R(M)$ values measured by two instruments decrease exponentially with M_n.

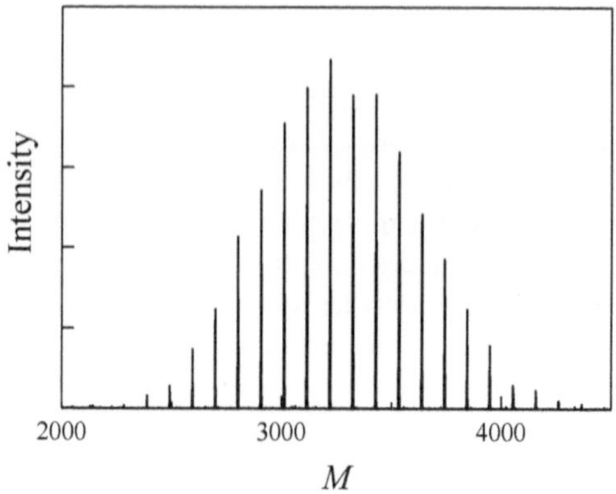

Fig. 2.18 Example of MALDI-TOF-MS spectrum for a polystyrene sample

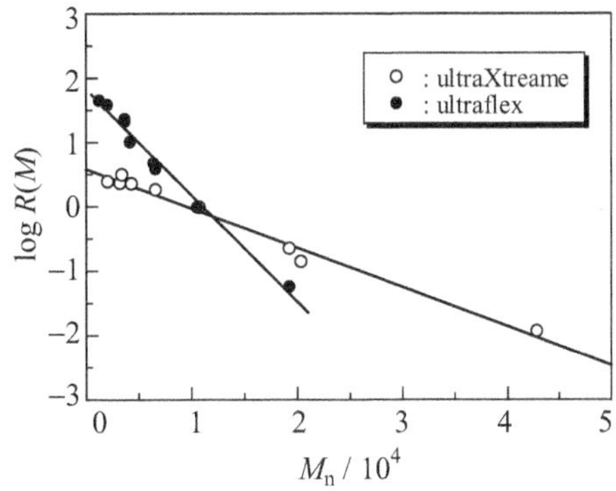

Fig. 2.19 Relative integrated intensity of MALDI-TOF-MS spectrum for polystyrene samples measured by two instruments

2.3.3 *Comparison of Molecular Weight Distributions Determined by Different Methods*

2.3.3.1 Comparison Between SEC-Calibration Curve and SEC-MALS Methods

Figure 2.20 shows the molecular weight for three standard polystyrene samples (F-128, F-80, F-40 with $M_w = 1.07 \times 10^6$, 6.86×10^5, and 3.43×10^5, respectively) determined by MALS detector plotted against the elution volume V, along with the peaks for the solute concentration detected by RI detector [11]. For these measurements, linear columns, containing gel beads with wide-range different pore sizes, were used. The calibration curve for such columns is expressed by the linear function of V as follows:

$$\log M = a_1 + a_2 V \tag{2.3.3}$$

Here, a_1 and a_2 are constants. In the figure, except for the value at the peak top, M_w values determined by MALS deviate from the calibration line of Eq. 2.3.3. This is because the solute diffused when it passed through the column, broadening the peak width, although the molecular weight distribution of each sample was very narrow.

To estimate the effect of peak-broadening and the real molecular weight distribution, the chromatogram for each sample is reproduced by calculation. The chromatogram can be expressed as the convolution of the weight fraction w_i of the ith component with molecular weight M_i and the peak-broadening function $G(V - V_i)$ as follows [12]:

$$c(V) = m_0 \sum_i w_i G(V - V_i) \tag{2.3.4}$$

Here, m_0 means the injected mass of the solute, and V_i is the elution volume for the ith component, which is expected to be related to M_i by Eq. 2.3.3. If the molecular weight distribution of the polymer samples is expressed by the log-normal

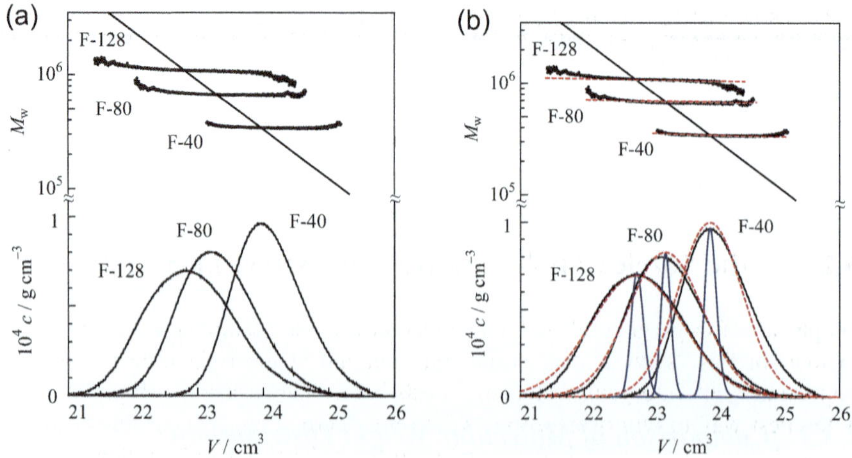

Fig. 2.20 Weight-average molecular weight and the solute mass concentration by SEC-MALS measurements for three polystyrene samples (**a**) and the same data with the calculated values (**b**) [11]

distribution and the M-V relation is given by Eq. 2.3.3, w_i is expressed by the Gaussian function as follows [12]:

$$w_i = \frac{1}{\sqrt{2\pi\sigma_{\mathrm{M}}^2}} \exp\left[-\frac{(x_i - x_{\mathrm{P}})^2}{2\sigma_{\mathrm{M}}^2}\right] \tag{2.3.5}$$

Here, x_i, x_{P}, and σ_{M}^2 denote the degree of polymerization of the ith component and that of the component at the peak top, and the variance for the molecular weight distribution, respectively. We may also assume the Gaussian function for $G(V - V_i)$:

$$G(V - V_i) = \frac{1}{\sqrt{2\pi\sigma_{\mathrm{B}}^2}} \exp\left[-\frac{(V - V_i)^2}{2\sigma_{\mathrm{B}}^2}\right] \tag{2.3.6}$$

Here, σ_{B} is the variance for the peak broadening. Based on these equations, the weight-average molecular weight of the sample eluted at V is calculated by the following equation:

$$M_{\mathrm{w}} = \sum_i M_i w_i G(V - V_i) / \sum_j w_j G(V - V_j) \tag{2.3.7}$$

The calculated results are shown in Fig. 2.20b, where dashed lines show the calculated values of M_{w} and c as functions of V from Eqs. 2.3.3 to 2.3.7 with the appropriate values for x_{P}, σ_{M}, and σ_{B}. The narrower peaks indicated by the solid lines show the ideal ones when there is no peak-broadening effect. The good agreements

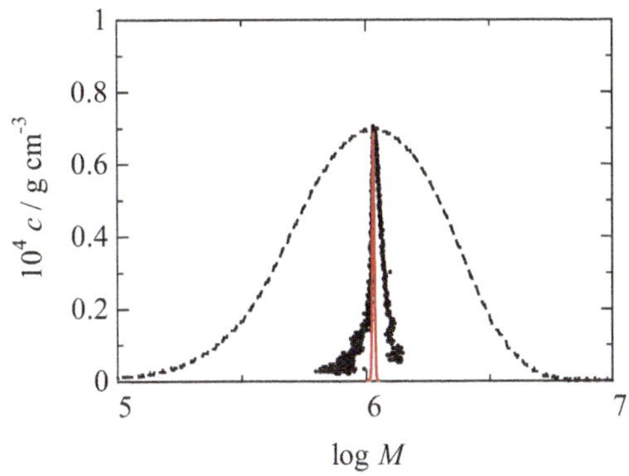

Fig. 2.21 Molecular weight distribution curves determined by SEC-calibration curve (dashed line), SEC-MALS (dots), and calculation (solid line) with Eq. 2.3.5 for F-128

between the experimental and calculated peaks show that the calculation process is accurate. The difference between the observed and calculated peak widths indicates the seriousness of the peak broadening effect in SEC.

Figure 2.21 shows the peaks plotted against log M, where the molecular weight distributions determined by SEC-calibration curve (dashed line), SEC-MALS (dots), and calculated (solid line) by Eq. 2.3.6 are compared. This figure shows the molecular weight distribution determined by SEC-calibration curve is much wider because of the peak-broadening effect. That determined by SEC-MALS is close to the calculated value although the distribution curve is still wider than the calculated one. We note that the calculated distribution curve is close to the Poisson distribution expected for samples synthesized by anionic polymerization.

2.3.3.2 Comparison Between SEC-Calibration Curve and MALDI-TOF-MS Methods

For low molecular weight samples, in addition to the above two methods, MALDI-TOF-MS measurements can be applied. Figure 2.22 shows the SEC chromatogram for a polystyrene sample with $M_w = 3.4 \times 10^3$ along with the calibration curve. The calibration curve is not straight since the SEC column used for this measurement is not of the linear type. Using this calibration curve, the bottom axis can be changed to the molecular weight to obtain the molecular weight distribution curve, as shown in Fig. 2.23, in which the peaks of the MALDI-TOF-MS measurement are also indicated. Since the original peak intensities of the latter measurement are proportional to the number of molecules, they are multiplied by the molecular weight in the figure. The peak heights are also corrected by the molecular weight dependence of the intensity, by using the relative intensity indicated in Fig. 2.19. The difference in the peak top positions between the two measurements may be due to the slight error in the calibration curve. The width of the molecular weight distribution observed by SEC is rather wider than that determined by MALDI-TOF-MS, reflecting the peak-broadening effect. The difference in the peak broadening effect

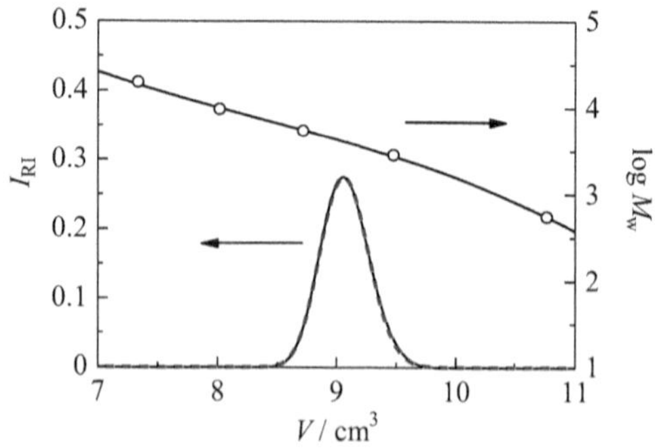

Fig. 2.22 SEC chromatogram for PS3400 ($M_w = 3.4 \times 10^3$, lower) and the calibration curve (upper). Calculated SEC curve is shown by the dashed line

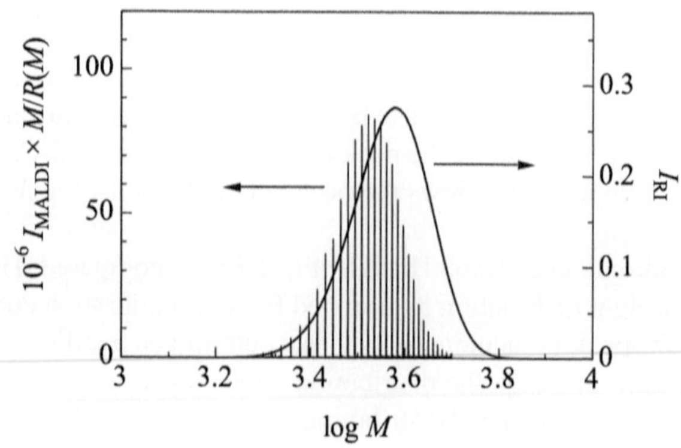

Fig. 2.23 Molecular weight distribution determined curves determined by SEC and MALDI-TOF-MS measurements

between Figs. 2.21, 2.22 and 2.23 can be attributed to the difference in the number of columns used, three columns for the former measurement and one column for the latter measurement. After adjusting the peak top position, the SEC chromatogram is calculated based on the molecular weight distribution obtained by the MALDI-TOF-MS measurement. The calculation process is the same as that described in the last section. The calculated values closely fit the experimental data, showing that the peak-broadening effect in SEC spreads the molecular weight distribution.

References

1. Flory PJ (1943) Molecular weights and intrinsic viscosities of polyisobutylenes. J Am Chem Soc 65:372–382. https://doi.org/10.1021/ja01243a022
2. Aota Y et al (2010) Band gap and absorption profile change by changing molecular weight and conformation of water-soluble narrow-band-gap polymers. Chem Lett 39:1288–1290. https://doi.org/10.1246/cl.2010.1288
3. Brandrup J, Immergut EH (eds) (1989) Polymer handbook 3rd ed, John Wiley & Sons, New York.
4. Einaga Y, Miyaki Y, Fujita H (1979) Intrinsic viscosity of polystyrene. J Polym Sci: Polym Phys Ed 17: 2103–2109. https://doi.org/10.1002/pol.1979.180171205

5. Flory PJ, Fox TG (1951) Treatment of Intrinsic Viscosities. J Am Chem Soc 73: 1904–1908. https://doi.org/10.1021/ja01149a002
6. Itou T, Chikiri H, Teramoto A, Aharoni SM (1988) Wormlike chain parameters of poly(hexyl isocyanate) in dilute solution. Polym J 20: 143–151. https://doi.org/10.1295/polymj.20.143
7. Grubisic Z, Rempp P, Benoit H (1967) A universal calibration for gel permeation chromatography. J Polym Sci Part B: Polym Lett 5: 753–759. https://doi.org/10.1002/pol.1967.110050903
8. Haney MA (1985) The differential viscometer. I. A new approach to the measurement of specific viscosities of polymer solutions. J Appl Polym Sci: 30 3023–3036. https://doi.org/10.1002/app.1985.070300724
9. Cotts PM, Guan Z, McCord E, McLain S (2000) Novel branching topology in polyethylenes as revealed by light scattering and ^{13}C NMR. Macromolecules 33: 6945–6952. https://doi.org/10.1021/ma000926r
10. Brandt H, Ehmann T, Otto M (2010) Toward prediction: using chemometrics for the optimization of sample preparation in MALDI-TOF MS of syntheticpolymers. Anal Chem 82: 8169–8175. https://doi.org/10.1021/ac101526w
11. Kondo Y, Nakamura Y (2012) Effects of molecular-weight distribution on SEC-MALS measurements. Kobunshi Ronbunsyu 69, 387–391. https://doi.org/10.1295/koron.69.387
12. Jackson C, Yau WW (1993) Computer simulation study of size-exclusion chromatography with simultaneous viscometry and light scattering measurements J Chromatgr 645: 209–217. https://doi.org/10.1016/0021-9673(93)83380-B

Chapter 3
Molecular Size and Shape

3.1 Overview

As mentioned in the previous chapter, the second virial coefficient A_2 and the weight-average molecular weight M_w are determined by light scattering measurements. The former quantity is obtained from the concentration dependence of the Rayleigh ratio R_θ at the scattering angle θ extrapolated to $\theta = 0$, R_0, and the latter quantity is from the extrapolation of R_0 to $c = 0$. Another quantity of the polymer dimension, the mean-square radius of gyration $\langle S^2 \rangle$ is obtained from the angular dependence of R_θ extrapolated to $c = 0$, $(R_\theta)_{c=0}$. When the molecule is modeled by an array of $(n + 1)$ beads of the same mass, this quantity is defined by the following equation:

$$\langle S^2 \rangle = \frac{1}{n+1} \sum_{i=0}^{n} \langle S_i^2 \rangle \tag{3.1.1}$$

Here, $\mathbf{S}_i$ is the vector from the center of mass of the molecule to the ith bead and $\langle \cdots \rangle$ means the statistical average. The sum is taken over all the beads. $\langle S^2 \rangle$ is obtained by the initial slope of the reciprocal $R_\theta$plotted against the square of the scattering vector, q^2. Since the dependence of R_θ on θ is specific to the molecular shape, the particle scattering factor $P(\theta) \equiv (R_\theta)_{c=0}/(R_0)_{c=0}$ is used to estimate the polymer structure. Another method to estimate the molecular structure is to analyze the molecular weight dependence of $\langle S^2 \rangle$. For example, $\langle S^2 \rangle$ for random coil chain is proportional to molecular weight M while that for rigid rod chain is proportional to M^2. Molecules having the middle property of these two chains are expressed by the wormlike chain model. This model is indicated by two parameters, the stiffness parameter, $\lambda^{-1}$and the contour length L of the chain. For the wormlike chain, λ^{-1} is twice of the persistence length l_p, which is defined by

Authors: K. Terao (terao.ken.sci@osaka-u.ac.jp) (Sect. 3.2) and Y. Nakamura (Sects. 3.1, 3.3, and 3.4).

Y. Nakamura et al., *Characterization of Polymers in Solution*,
https://doi.org/10.1007/978-981-96-6677-5_3

$$l_{\text{p}} = \lim_{L \to \infty} \langle \boldsymbol{R} \cdot \boldsymbol{u}_0 \rangle \tag{3.1.2}$$

Here, **R** and $\mathbf{u}_0$ are the end-to-end vector and the unit tangent vector at one of the chain ends, respectively. L is related to the molecular weight M by $L = M/M_{\text{L}}$ with the molecular weight per unit contour length M_{L}. By fitting the experimental $\langle S^2 \rangle$ plotted double-logarithmically against M, λ^{-1} (or l_{p}) and M_{L} are determined. Similar analyses are made for the intrinsic viscosity $[\eta]$ and the hydrodynamic radius R_{H}. Details of the data analyses are discussed in Sect. 3.4.1.

3.2 Radius of Gyration and Scattering Function

3.2.1 Light Scattering

Light is scattered in the air. The sky thus looks blue. This indicates that some fluctuation exists in the air. This is because of the Rayleigh scattering of visible light. The angular distribution of scattered light reflects the size and/or shape of scatterer when the structure is similar size range of the wavelength of the light. Indeed, scattering behavior of sunlight can be used to analyze the interstellar materials. Since this scattering phenomenon is also found for dilute solutions of polymers, it is an indispensable method to characterize polymeric materials in solution. The size range of polymers are usually similar or smaller than that for the visible light. Thus, this method is applicable to observe not only molar mass but also dimensional properties and intermolecular interactions of polymers. More recently, this method is used as a detector of the high-performance liquid-chromatography (HPLC) system to characterize polydisperse materials. In this chapter, we summarize the basic theoretical background, experimental procedure, and data analysis of static light scattering method.

3.2.1.1 Scattering from Small Polymer Molecules

Since the visible light is an electromagnetic wave, a vibrating dipole is created with the frequency of the light. The dipole radiates the light wave for which the wavelength and phase are identical to the incident light. This is the elastic-light scattering phenomenon. According to the electromagnetics, the magnitude of the induced dipole is proportional to the polarizability α of the particle (or polymer molecules). Now, we assume that there are ν polymer molecules in the scattering volume V. The scattering intensity I observed at the detector for which the distance from the molecules is r is thus written as

$$I = \frac{I_0}{r^2} \left(\frac{\alpha}{4\pi\varepsilon_0} \right)^2 \left(\frac{2\pi}{\lambda_0} \right)^4 \nu \tag{3.2.1}$$

where λ_0, I_0, and ε_0 are the wavelength in a vacuum, the intensity of the incident light, and the permittivity of vacuum. In general, the scattering intensity from polymer solutions includes scattered light from the solvent, that is, the density fluctuation. Therefore, the excess scattering intensity ΔI from the polymer molecules can be obtained by subtracting the scattering intensity of the pure solvent from that of the solution. This ΔI is proportional to the square of the excess polarizability $\Delta\alpha$, which is written as

$$\Delta I = \frac{I_0}{r^2}\left(\frac{\Delta\alpha}{4\pi\varepsilon_0}\right)^2\left(\frac{2\pi}{\lambda_0}\right)^4 \nu \tag{3.2.2}$$

According to the electromagnetics, $\Delta\alpha$ can be written at the limit of infinite dilution as

$$\frac{\Delta\alpha}{4\pi\varepsilon_0} = \frac{n_0 M}{2\pi N_A}\left(\frac{\partial n}{\partial c}\right) \tag{3.2.3}$$

Here, n_0, M, N_A, and $\partial n/\partial c$ are the refractive index of the solvent, the molar mass of the polymer, the Avogadro number, and the excess refractive index increment, respectively. If we introduce the Rayleigh ratio R_q $(= R_\theta)$ defined as $R_q \equiv \Delta I r^2/I_0 V$, Eq. 3.2.2 can be deformed as

$$R_q = \left(\frac{\Delta\alpha}{4\pi\varepsilon_0}\right)^2\left(\frac{2\pi}{\lambda_0}\right)^4 \frac{\nu}{V} = \left(\frac{\Delta\alpha}{4\pi\varepsilon_0}\right)^2\left(\frac{2\pi}{\lambda_0}\right)^4 \frac{N_A c}{M} \tag{3.2.4}$$

taking into $c = M\nu/N_A V$ account. Substituting Eq. 3.2.3 into Eq. 3.2.4, the following expression is obtained

$$R_q = KMc \tag{3.2.5}$$

where K is the optical constant defined by Eq. 2.2.17. Since K can be evaluated experimentally, M is determined from the excess scattering intensity.

3.2.1.2 Scattering from Relatively Large Polymer Molecules

When the size of the dispersed molecules is larger than 1/20 of the wavelength, intramolecular interference may not be ignorable. Since this effect significantly depends on the scattering angle θ, the scattering intensity also depends on θ. This is schematically represented in Fig. 3.1. Let us consider two incident light waves with exactly the same phase. The first one is scattered at the point i while the other is scattered at j. As seen from the figure, the first one reaches the detector slightly earlier than the other. Thus, the intensity of the combined wave is slightly smaller than the twice the single wave. The path-length difference can be written as $(\mathbf{e}_0 - \mathbf{e}) \cdot \mathbf{R}_{ij}$ where $\mathbf{e}_0$ and $\mathbf{e}$ denote the unit vectors of the incident and scattered light,

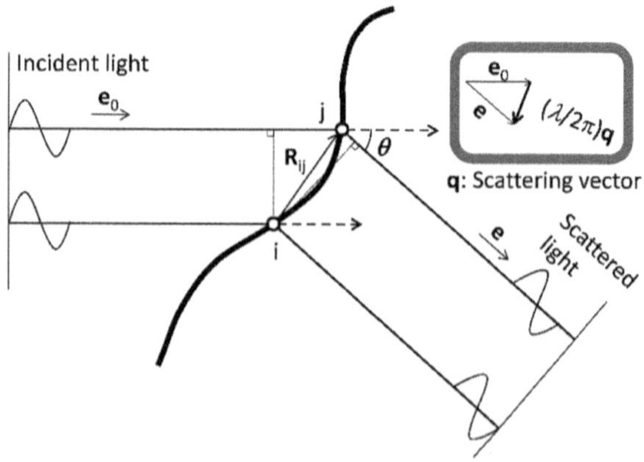

Fig. 3.1 Interference effects on the scattered light

respectively, and $\mathbf{R}_{ij}$ is the vector from i to j. Thus, the resulting phase difference can be written as

$$\frac{2\pi}{\lambda}(\mathbf{e}_0 - \mathbf{e}) \cdot \mathbf{R}_{ij} = -\mathbf{q} \cdot \mathbf{R}_{ij} \tag{3.2.6}$$

where λ is the wavelength in the medium and equal to λ_0/n_0. The scattering vector $\mathbf{q}$ on the right-hand side is defined as $(2\pi/\lambda)(\mathbf{e}_0 - \mathbf{e})$. The magnitude q of $\mathbf{q}$ is thus written as

$$q = \frac{4\pi}{\lambda}\sin\frac{\theta}{2} = \frac{4\pi n_0}{\lambda_0}\sin\frac{\theta}{2} \tag{3.2.7}$$

Considering the phase difference from the two points, the scattering intensity can be calculated as follows:

$$R_q = KMcP(q) \tag{3.2.8}$$

where $P(q)$ denotes the particle scattering function. Since the intramolecular interference is negligible at $\theta = 0$, $P(q)$ at $\theta = 0$ is unity. Consequently, the molar mass can be determined from the extrapolated value of R_q/KMc to $q = 0$ when the concentration of the polymer is low enough to ignore intermolecular interference as described below. The particle scattering function $P(q)$ defined as the ratio of the scattering intensity at q to that at $q = 0$ reflects the size and shape of the polymer molecules. When we consider each polymer molecule is composed of n scatterers and randomly oriented, $P(q)$ can be written as

$$P(q) \equiv \lim_{c\to 0}\left(\frac{R_q}{R_0}\right) = \frac{1}{n^2}\sum_{i=1}^{n}\sum_{j=1}^{n}\left\langle\frac{\sin qr_{ij}}{qr_{ij}}\right\rangle \tag{3.2.9}$$

where $\langle \cdots \rangle$ and r_{ij} denote the statistical average and the magnitude of $\mathbf{R}_{ij}$, respectively. When we expand Eq. 3.2.9 around $q = 0$, $P(q)$ is related to the radius of gyration $\langle S^2 \rangle$ as

$$P(q) = 1 - \frac{1}{6N^2} \sum_{i=1}^{N} \sum_{j=1}^{N} \langle r_{ij}^2 \rangle q^2 + \cdots = 1 - \frac{1}{3} \langle S^2 \rangle q^2 + \cdots \tag{3.2.10}$$

This equation is valid for any shaped particles (or polymer molecules) in solution when the scattering electric field from each part of the particle is the same. If we consider the center of gravity of each particle is easily calculatable, such as spheres and rigid rods, the following expression can also be used instead of Eq. 3.2.9 to calculate $P(q)$ theoretically.

$$P(q) = \frac{1}{N} \left\langle \left[\sum_{i=1}^{N} \frac{\sin(qS_i)}{qS_i} \right]^2 \right\rangle \tag{3.2.11}$$

3.2.1.3 Scattering from Multiple Polymer Molecules in Solution

When the concentration of the polymer is not exactly dilute, intermolecular excluded-volume effect may significantly cause the scattering intensity even at low q range. Considering this effect, Eq. 3.2.8 can be rewritten as

$$R_q = KMcP(q)S(q) \tag{3.2.12}$$

where $S(q)$ is defined as

$$S(q) \equiv \frac{R_q/Kc}{\lim\limits_{c \to 0} (R_q/Kc)} = 1 + \nu \left\langle \frac{\sin qR_{kl}}{qR_{kl}} \right\rangle \tag{3.2.13}$$

where R_{kl} denotes the distance between the barycenters of the two particles. To calculate $S(q)$ theoretically, both the shape of each polymer molecule and intermolecular interactions must be considered. Meanwhile, at the limit of $q = 0$, concentration dependence of the scattering intensity can be treated as only thermodynamically. According to the Einstein-Smoluchowski theory, when the intraparticle interference is negligible, that is, $P(q) = 1$, Kc/R_q at $q = 0$ can be expressed as

$$\frac{Kc}{R_0} = \frac{\partial}{\partial c} \left(\frac{\Pi}{RT} \right) \tag{3.2.14}$$

where Π is the osmotic pressure. Substituting the virial expansion of Π, Eq. 2.2.2, we obtain

$$\frac{Kc}{R_0} = \frac{\partial}{\partial c}\left(\frac{\Pi}{RT}\right) = \frac{1}{M} + 2A_2c + 3A_3c^2 + \cdots \tag{3.2.15}$$

This equation indicates that the second and third virial coefficients, A_2 and A_3, can be evaluated from the concentration dependence of Kc/R_0.

As is well known, molar mass of many polymer samples is not uniform. In such case, the obtained molar mass from Eq. 3.2.5 is a certain average of the molar masses. The Rayleigh ratio for the polydisperse system can be written with the weight fraction w_s, the molar mass M_s, and the particle scattering function $P_s(q)$ of each component as

$$\left(\frac{R_q}{Kc}\right)_{c=0} = \sum_{s=1}^{r} w_s M_s \frac{\sum_{s=1}^{r} w_s M_s P_s(q)}{\sum_{s=1}^{r} w_s M_s} = M_w P_z(q) \tag{3.2.16}$$

where M_w and $P_z(q)$ are the weight-average molar mass and the z-average particle scattering function, respectively. The latter can be expanded around $q = 0$ as

$$P_z(q) = \frac{\sum_{s=1}^{r} w_s M_s \left(1 + \frac{1}{3}\langle S^2\rangle_s q^2 + \cdots\right)}{\sum_{s=1}^{r} w_s M_s} = 1 - \frac{1}{3}\frac{\sum_{s=1}^{r} w_s M_s \langle S^2\rangle_s}{\sum_{s=1}^{r} w_s M_s} q^2 + \cdots$$
$$= 1 - \frac{1}{3}\langle S^2\rangle_z q^2 \tag{3.2.17}$$

Here, $\langle S^2\rangle_z$ is the z-average mean-square radius of gyration. The second virial coefficient for the polydisperse samples is somewhat different from that determined by the osmotic pressure while those for the monodisperse sample are essentially the same.

3.2.2 *Small-Angle X-ray and Neutron Scattering*

The above-mentioned method for light scattering is also applicable when we utilize hard X-ray or cold neutron as the incident light. These methods cover a wider q range whereas the maximum q value for the light scattering method is at most 0.03 nm^{-1}: The minimum $\langle S^2\rangle$value from light scattering is about 100 nm^2. On the contrary, the methods with X-ray and neutron are applicable for smaller particles, and hence, local structural information can also be available. In actuality, substantially the same data treatment is available for these experiments if we determine R_q and K from the adequate procedure. We note that the scattering differential cross section is frequently used instead of the Rayleigh ratio. The following shows the major characteristic using the two quantum beams.

3.2.2.1 Small-Angle X-ray Scattering (SAXS)

To attain the adequate transparency of X-ray, the photon energy should be 8 keV or higher, corresponding to $\lambda_0 < 0.15$ nm. This value is at least 1/2500 shorter comparing to the visible light, the scattering data at the very low θ region should be used for polymeric materials to observe the scattering data at an adequate q range. In general, transparency of the X-ray becomes better with lowering the wavelength while the q value at the same θ increases. Thus, even though the use of synchrotron radiation, the photon energy is at most 30 keV. The reduced scattering intensity R_q of X-ray for which the definition is the same as that for visible light can be written as

$$R_q = K_X Mc \tag{3.2.18}$$

$$K_X = N_A I_e \Delta z^2 \tag{3.2.19}$$

where I_e is the Thomson factor 7.9×10^{-26} cm^2 (or square of the classical electron radius) and Δz is the X-ray contrast factor defined by

$$\Delta z = z - \overline{v}\rho_{e,s} \tag{3.2.20}$$

Here, z is the number of moles of electrons per unit mass of the solute, $\overline{v}$ is the partial specific volume of the solute, and $\rho_{e,s}$ is the electron density of the solvent. Therefore, the contrast factor is essentially different from those for the light scattering. For example, SAXS is suitable for fluoropolymers in solution because they cannot be characterized by light scattering because of their low refractive index. Another important point is that this method is applicable for the colored polymers including fluorescent polymers. On the contrary, absorption of the X-ray by the solute and solvent cannot usually be ignored for X-ray. The quantitative measurement of the Rayleigh ratio is generally more difficult comparing with the light scattering method. Especially, SAXS cannot be applied for some solvent systems including relatively heavy atoms, such as chlorine and sulfur because of their low transparency of X-ray.

3.2.2.2 Small-Angle Neutron Scattering (SANS)

Cold neutron beam obtained from the atomic reactor or the proton accelerator is also useful to measure elastic scattering from the polymeric materials in solution. Since the contrast factor is determined by the nature of nucleus consisting of the particle, light element including hydrogen is rather emphasized. Indeed, the contrast difference between hydrogen and deuterium is detectable. See the webpage of National Institute of Standards and Technology: NIST for the scattering length of each element. For the ordinal polymeric materials, deuteride solvent or polymers are therefore used to attain enough scattering intensity. Furthermore, if we choose a certain mixture of hydride and deuteride solvents and the resultant solvent has the same scattering constant for a dissolved substance, no scattering can be observed. For the solution including two or more solutes, it is a powerful tool to observe each substance in the solution. This is called as the contrast matching method.

3.2.3 *Measurements and Data Analysis*

3.2.3.1 Light Scattering Measurements

Scattering intensity is frequently measured for the static light scattering measurement to determine the above-mentioned physical properties, that is, M, A_2, and $\langle S^2 \rangle$. Specially designed light scattering photometer with cylindrical cells made of glass are used to determine the scattering intensity as a function of the scattering angle θ. Each cell is immersed in an optically matching bath in which the refractive index n_o is close to the glass cell to avoid stray light. Various visible laser can be used for the incident light. Vertically polarized light is usually used when the optical anisotropy of the polymer is insignificant (see below). When we do not use any polarizers at the detector, we call it as the Uv arrangement. We can add polarizing prism and/or interference filters to the detector if needed.

Since appropriate slit is equipped for the detector, scattering light from a specific volume, that is, the scattering volume, is detected. As shown in Fig. 3.2, scattering volume is proportional to the area of the quadrangle. When we assume the cross section of the incident light is a foursquare with a length of a and both the slit width and height at the detector are b and $b < a$, the scattering volume V can be calculated to be $ab^2 \sin\theta$ when the refractive index n of the test solution is equivalent to n_o. Thus, V depends on the scattering angle θ and it can be compensated by multiplying $1/\sin\theta$. If n is lower than n_o, the scattering volume is larger than the original V. It can be written as

$$V = \left(\frac{n_o}{n}\right)^2 \frac{ab^2}{\sin\theta} \tag{3.2.21}$$

It is noted that the scattering volume should be calculated by an appropriate equation for different shaped light scattering cells, such as a flow cell for a multi-angle light-scattering detector for HPLC system.

The Rayleigh ratio cannot usually be determined with the light scattering photometers. The instrument constant can be estimated from the scattering intensity of a referenced material. Table 3.1 shows the Rayleigh ratio R_{90} for toluene and benzene at $\theta = 90°$ and at λ_0. Utilizing these values, we can determine R_q for the test solution with the excess scattering intensity ΔI as

$$R_q = \left(\frac{n}{n_s}\right)^2 \sin\theta \, \Delta I \tag{3.2.22}$$

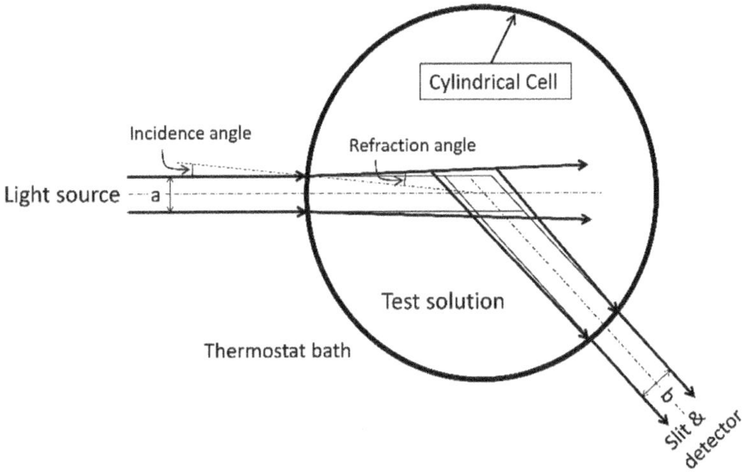

Fig. 3.2 Schematic representation of light scattering cell and the scattering volume

Table 3.1 Rayleigh ratio of the standard materials at 25 °C (Uv arrangement)

λ_0/nm	$R_{90}/10^{-6}$ cm^{-1}	
	Benzene	Toluene
436	65.9	
488	34.1	39.1
532		27.2
546	23.0	
633	11.8	13.6

with n_s being the refractive index of the standard material.

Since the scattering intensity is proportional to Mc, a small amount of larger dusts than the objective polymers cause significant error. Optical purification is thus necessary to measure light scattering intensity with high accuracy. Filtration with an appropriate membrane filter and/or centrifugation of solutions are effective to remove the dusts from the solution. Passing through size-exclusion chromatography (SEC) columns is also a very powerful method for the purification. Thus, recent development of SEC-light scattering is one of the easiest methods to determine light scattering intensity of dilute polymer solutions.

3.2.3.2 SAXS and SANS Measurements

The low scattering angle usually at most 5° is important both for SAXS and SANS measurements to acquire the scattering data of the relevant q range. The flat plate cell is preferably used for the methods. Because of the very small refractive indices of X-ray and neutron beam, cylindrical cell can also be utilized. The scattered X-ray or neutron are detected by using the two-dimensional detector. A circular average procedure is used to determine the scattering intensity as a function of θ or q. For SAXS measurements, the window materials of the cell must be thin to retain the transparence, and furthermore, the path length of X-ray is set to be 1–3 mm because the absorption is significant comparing with visible light.

3.2.3.3 Data Analysis

As is shown in the previous sections, double extrapolation of Kc/R_q both to $q = 0$ and $c = 0$ is an important step to determine the molar mass M, intermolecular interactions, and dimensional properties. Appropriate extrapolation procedures are needed to obtain the above-mentioned physical properties with an adequate accuracy. For example, the Zimm plot, the square-root Zimm plot (Berry plot), and the Guinier plot are frequently used to analyze the scattering data.

The following equations are proposed at infinite dilution to extrapolate the scattering data to $q = 0$.

$$\left(\frac{Kc}{R_q}\right)_{c=0} = \frac{1}{M}\left(1 + \frac{1}{3}\langle S^2\rangle q^2 + \cdots\right) \quad \text{(Zimm plot)} \tag{3.2.23}$$

$$\left(\frac{Kc}{R_q}\right)^{1/2}_{c=0} = \frac{1}{M^{1/2}}\left(1 + \frac{1}{6}\langle S^2\rangle q^2 + \cdots\right) \quad \text{(Berry plot)} \tag{3.2.24}$$

$$\left[\ln\left(\frac{R_q}{Kc}\right)\right]_{c=0} = \ln M - \frac{1}{3}\langle S^2\rangle q^2 + \cdots \quad \text{(Guinier plot)} \tag{3.2.25}$$

These plots are also feasible to estimate R_0 at a finite c. Angular dependence of $P(q)$ can be calculated for the actual particles and/or the polymers by using Eqs. 3.2.9 and 3.2.11. The resulting expressions for various shaped particles and polymer chains are summarized in Table 3.2. The three plots, Zimm, Berry, and Guinier, are illustrated in Fig. 3.3 for the four representative shapes, that is, sphere, ring, Gaussian chain, and thin rod, are shown in each panel to compare the linearity at low q region. The initial slope is independent of the shape as explained with Eq. 3.2.10. The Guinier plot is suitable for spheres and rings while the Berry plot is applicable for the Gaussian chains and rigid rods. When we consider the polydispersity, the Zimm plot is rather suitable for rigid rods.

The intermolecular interactions in terms of the virial coefficients are effectible to the scattering intensity at finite concentration. As in the case of the q dependence, three plots, Zimm, Berry, and Guinier, can be utilized to analyze the concentration dependence Kc/R_q at $q = 0$. As shown in Fig. 3.4, The Zimm plot, Kc/R_q versus c, is suitable to estimated M and A_2 when the higher order virial coefficients are not appreciable. In contrast, it is known that the following square-root Zimm plot derived from Eq. 3.2.15 gives adequate accuracy for the Gaussian chains.

Table 3.2 Particle scattering functions and gyration radii for various scattering bodies

Scattering body	$P(q)$	$\langle S^2 \rangle$
Sphere with the diameter R	$\left(\frac{3}{x^3}\right)^2 (\sin x - x\cos x)^2,\ x \equiv qR$	$\frac{3}{5}R^2$
Spherical shell with the diameter R	$\left(\frac{\sin x}{x}\right)^2$	R^2
Disk with the diameter R[1]	$\frac{2}{x^2}\left(1 - \frac{J_1(2x)}{x}\right)$	$\frac{1}{2}R^2$
Circle with the diameter R[2]	$\frac{\pi J_1(2x)H_0(2x)+J_0(2x)[2-\pi H_1(2x)]}{2}$	R^2
Thin rod with the length L[3]	$\frac{2}{x^2}\left[y\ \mathrm{Si}(y) + \cos y - 1\right],\ y \equiv qL$	$\frac{1}{12}L^2$
Cylinder with the length L and diameter R[4]	$\frac{2}{x^2}\left[y\mathrm{Si}(y) + \cos y - 1\right]\left[\frac{2J_1(x)}{x}\right]^2$	$\frac{1}{12}L^2 + \frac{1}{2}R^2$
Gaussian chain[5]	$\frac{2}{z^2}(e^{-z} - 1 + z),\ z \equiv \frac{Nb^2}{6}q^2$	$\frac{1}{6}Nb^2$

[1]$J_\alpha(x)$: Bessel function of the first kind. [2]$H_\alpha(x)$: Struve function. [3]$\mathrm{Si}(x)$: Sine integral. [4]Approximate expression. See Ref. [1]. [5]N: bond number, b: bond length

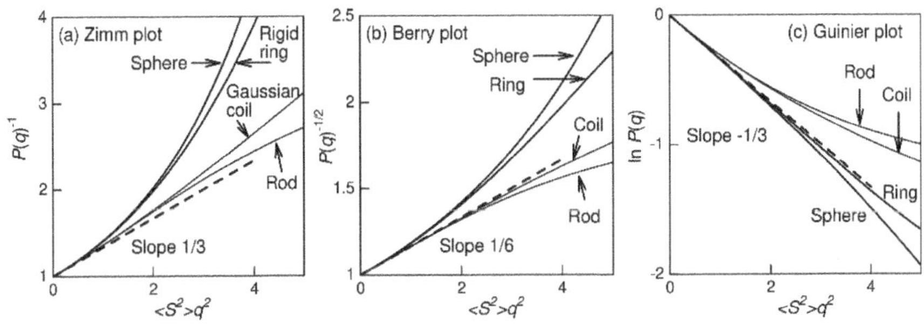

Fig. 3.3 Angular dependence of scattering intensity. **a** Zimm plot, **b** Berry plot, and **c** Guinier plot for the various scattering bodies

$$\left(\frac{Kc}{R_0}\right)^{1/2} = \frac{1}{M^{1/2}}\left[1 + A_2Mc + \frac{3}{2}A_2^2M^2\left(g - \frac{1}{3}\right)c^2 + \cdots\right] \quad (3.2.26)$$

Here, g is the reduced third virial coefficient defined by Eq. 2.2.7.

Table 3.3 summarizes the g values for the various scattering bodies in solution. The Berry plot is preferable for the Gaussian chains for which g is close to 1/3. For the spherical particles, the Guinier plots

$$\ln\left(\frac{R_0}{Kc}\right) = \ln M - 2A_2Mc - 3A_2^2M^2\left(g - \frac{2}{3}\right)c^2 + \cdots \quad (3.2.27)$$

are usually utilized owing to the larger g value. Table 3.4 shows the suitable plots to analyze both the angular and concentration dependence to obtain the accurate extrapolated values. Of course, if the sample has very large dispersity index, it affects

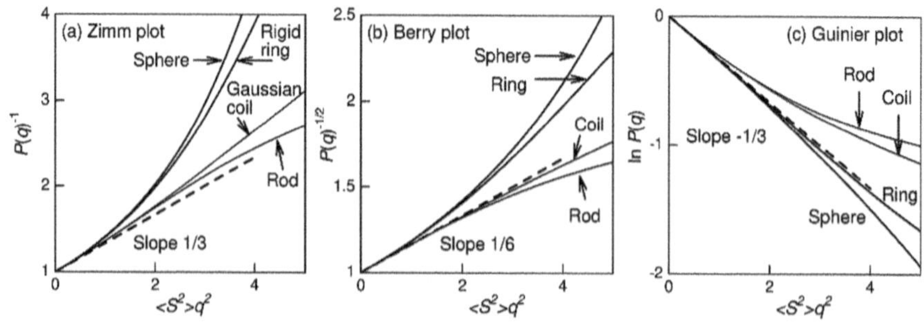

Fig. 3.4 Concentration dependence of scattering intensity. **a** Zimm plot, **b** Berry plot, and **c** Guinier plot for various scattering bodies

Table 3.3 Reduced third virial coefficients for various scattering bodies

Scattering body	g
Sphere with the diameter R	5/8
Cylinder with the length L and diameter R	$10d/3L$
Gaussian chain[1]	0.3–0.5

[1] N bond number, b bond length

Table 3.4 Recommended plots for the following scattering bodies

Scattering body	Concentration dependence	Angular dependence
Sphere	Berry or Guinier	Guinier
Rod	Zimm	Berry
Gaussian chain	Berry	Berry

the shape of the plot. Relatively narrow dispersed samples are desirable to determine the physical properties of polymer chains accurately.

Figure 3.5 illustrates an example of the actual analysis of the light scattering data for a polysilane in isooctane [2]. The Berry plot has good linearity to determine the weight-average molar mass M_w, A_2, and $\langle S^2\rangle_z$ on the basis of Eqs. 3.2.24 and 3.2.26. The resulting values of M_w, A_2, and $\langle S^2\rangle_z^{1/2}$ are 873 kg mol^{-1}, 1.5 × 10^{-4} m^3 mol kg^{-2}, 53.2 nm, respectively.

As shown in Table 3.2, plots of $P(q)$ versus q significantly depend on the shape of polymer molecules at high q region. Figure 3.6a, b displays the double logarithmic plots of $P(q)$ versus q for amylose tris(n-butylcarbamate) (ATBC) in tetrahydrofuran (THF) and methanol (MeOH) [3]. This polymer behaves as rigid rod in THF and semiflexible chain in MeOH. A peak is found at about 4 nm^{-1} in THF whereas it cannot be found in MeOH. This is most likely because of the rigid helical main chain in THF. However, the difference in the chain stiffness is insignificant between the two plots. The Holtzer plot [$qP(q)$ vs q] or the Kratky plot [$q^2P(q)$ vs q] are preferably

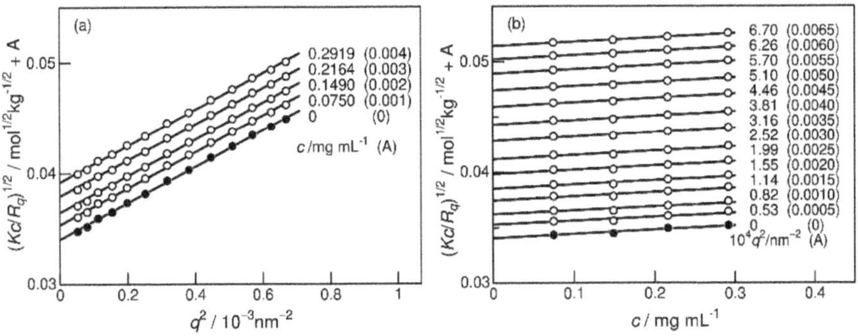

Fig. 3.5 Plots of $(Kc/R_q)^{1/2}$ versus q^2 at indicated c (**a**) and $(Kc/R_q)^{1/2}$ versus c at fixed q (**b**) for a polysilane sample in isooctane at 25 °C [2]. For clarity, the ordinate values of $(Kc/R_q)^{1/2}$ are shifted by A

used to emphasize the global conformational characteristics. The Holtzer plots for ATBC in the two solvents are shown in Fig. 3.6c, d. They clearly indicate that the difference in the chain stiffness at $q = 0.2\ \text{nm}^{-1}$: A peak is found only for semiflexible chain in MeOH. To analyze the data, the latest theory [4, 5] for the wormlike chain model is useful. The solid curves in the figures are the theoretical values for the wormlike chain. Alternatively, Fig. 3.6c, d show mostly flat region so-called the Holtzer plateau. For the thin polymer chains, the height of the plateau is equal to π/L with L being the contour length of the polymer. Gradually decreasing behavior at high q region is owing to the chain thickness effect. The following conventional equation can be used to estimate the chain thickness as [6]

$$P(q) = P_0(q)\exp[-(q^2d^2)/16] \tag{3.2.28}$$

with $P_0(q)$ and d being the particle scattering function for the thin chain and the chain diameter, respectively. Since $qP_0(q)$ reaches π/L as described above, plots of $\ln[qP_0(q)]$ versus q^2 in a certain q range obey a straight line for which the slope is $-d^2/16$. Thus, the plot is called as the cross-sectional plot. From the data in Fig. 3.6, the d value can be estimated from the slope in Fig. 3.7 to be 3.5 nm in THF and MeOH [7]. It is, however, noticed that the scattering data in the q range may be influenced by the electron density profile of the chain contour and solvent molecules nearby the polymer chain. If we choose ethyl lactate (EL) as a solvent, the cross-sectional plot has positive slope, and hence, the negative d^2 value. This behavior can be explained by the concentric double cylinder [8]. Furthermore, the relevant q range to apply the cross-sectional plot depends on the polymer and solvent. Therefore, direct comparison between the experimental data and theoretical values for the appropriate polymer chain model including the cylindrical wormlike chain model [4, 9] is strongly recommended.

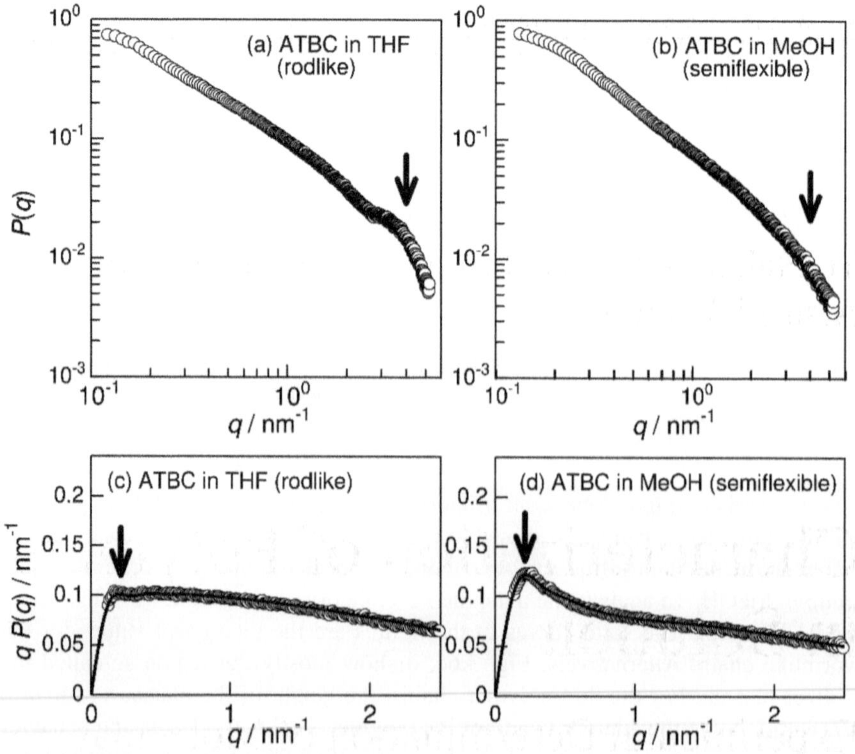

Fig. 3.6 Double logarithmic plots for $P(q)$ versus q for an amylose derivative sample in THF (**a**) and in MeOH (**b**), and the Holtzer plots for the same sample in THF (**c**) and in MeOH (**d**) [3]

Fig. 3.7 Cross-sectional plots for an ATBC sample in MeOH (circles), in THF (squares), and in ethyl lactate (triangles) at 25 °C [7]

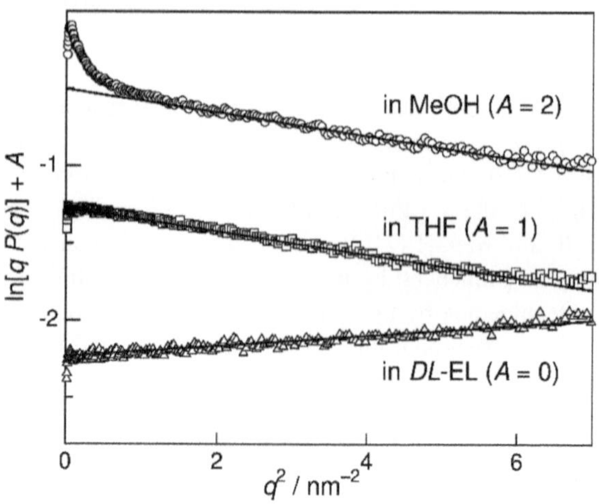

3.2.3.4 Notes on Actual Measurements

Optical anisotropy for light scattering measurement: When the anisotropy of the polarizability of the polymer molecules, Eq. 3.2.1 is no longer valid. In this case, we need the following two measurements. (1) A vertically oriented polarizer is set between the cell and the detector (Vv arrangement). (2) A horizontally oriented polarizer is located (Hv arrangement). If we designated the Rayleigh ratios $R_{q,\mathrm{Vv}}$ and $R_{q,\mathrm{Hv}}$ for the Vv and Hv arrangements, respectively, and they are extrapolated to $\theta = 0$ to determine $R_{0,\mathrm{Vv}}$ and $R_{0,\mathrm{Hv}}$, the parameter δ regarding the optical anisotropy can be determined with the following equation.

$$\left(\frac{R_{0,\mathrm{Hv}}}{R_{0,\mathrm{Vv}}}\right)_{c\to 0} = \frac{3\delta}{1+4\delta} \tag{3.2.29}$$

We can also use $R_{0,\mathrm{Uv}}$ for the Uv arrangement as

$$\left(\frac{R_{0,\mathrm{Hv}}}{R_{0,\mathrm{Uv}}}\right)_{c\to 0} = \frac{3\delta}{1+7\delta} \tag{3.2.30}$$

The apparent molar mass M_{app} estimated from Eq. 3.2.5 without considering the optical anisotropy can be larger than the true value as follows.

$$M_{\mathrm{app}} = (1+7\delta)M \tag{3.2.31}$$

Similar correction is also mandatory for A_2,

$$A_{2,\mathrm{app}} = (1+7\delta)^{-2}A_2 \tag{3.2.32}$$

These terms may be negligible when $\delta < 1/1000$. Of course, the optical anisotropy also affects $\langle S^2\rangle$ [10, 11]. The shape of the scattering body is mandatory to correct the effects on $\langle S^2\rangle$. Incidentally, this δ value for most polymers decreases with increasing molar mass. Figure 3.8 is an example for the M_{w} dependence of δ for amylose tris(3,5-dimethylphenylcarbamate) (ADMPC) in three different solvents. Considering that this effect is insignificant for SAXS and SANS and these methods are suitable to determine $\langle S^2\rangle$ for low molar mass polymer samples, not only light scattering but also SAXS and SANS are useful to determine the chain conformation of polymers accurately even though the optical anisotropy cannot be ignored.

Visible light absorption and fluorescence: Transparence of the incident light may cause a significant error for the scattering data. The apparent Rayleigh ratio $R_{q,\mathrm{app}}$ for the colored solution can be smaller than the actual R_q. It can be related to be

$$R_q = R_{q,\mathrm{app}} 10^{\gamma lc} \tag{3.2.33}$$

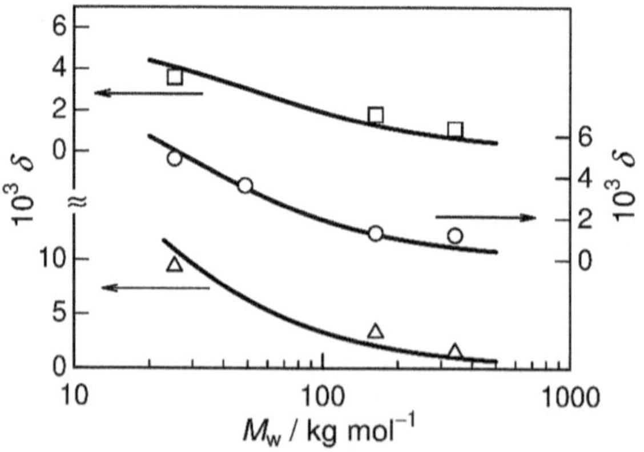

Fig. 3.8 Molar mass dependence of the optical anisotropy factor δ for ADMPC in MIBK (squares), MEK (circles), and MEA (triangles) at 25 °C. The curve represents the theoretical values calculated from the wormlike chain model. Reprinted with permission from Ref. [2]. Copyright (2010) American Chemical Society

where γ and l are the extinction coefficient and the diameter of the cell, respectively; note that the path length of the scattered light is the same as the diameter of the cylindrical cell. This effect mainly overestimates A_2.

However, if the polymer sample is fluorescent and/or the solution includes some fluorescent impurity, the observed scattering intensity should be larger than the actual value. An interference filter is effective to remove the fluorescence since it has longer wavelength than that for the incident light. Red laser is also a good choice for the light scattering measurement because both the absorption and fluorescence problems tend to diminish for the most polymers at the wavelength.

Scattering intensity and the virial coefficients: The incident light slightly attenuates by the light scattering phenomena while Eq. 3.2.1 ignores this effect, that is, called as the Rayleigh-Gans approximation. This is usually valid to determine molar masses and gyration radii. The second virial coefficient can be affected by the slight attenuation because it is determined by the concentration dependence of the scattering intensity. The Rayleigh ratio ignoring the slight attenuation by the light scattering is designated to be $R_{q,\mathrm{app}}$, which is related to R_q as

$$R_q = R_{q,\mathrm{app}}\exp(\tau l) \tag{3.2.34}$$

where τ is the turbidity. When the angular dependence of R_q is negligible, it can be written as

$$\tau = \frac{8\pi R_q}{3} \tag{3.2.35}$$

for the Uv arrangement. Furthermore, when this effect is not negligible, multiple scattering should also be considered. However, it is hard to estimate correctly because it may depend on the optical system including shape and/or size of the cell. It is thus desirable to choose longer wavelength and/or shorter path length of the light to diminish this effect.

Mixed solvent systems: When we choose a mixed solvent for the light scattering measurement, preferential adsorption should be considered. If we do not consider this effect, wrong molar mass and virial coefficient will be estimated. One of the applicable methods is that the specific refractive index increment $(\partial n/\partial c)_\mu$ at the fixed chemical potential is used instead of $\partial n/\partial c$. This can be determined by the same measurement for the above mentioned $\partial n/\partial c$, but each test solution is dialyzed by the mixed solvent prior to use. This method causes however somewhat error for A_2. To determine it accurately, light scattering measurements should be made for dialyzed solutions.

Light scattering photometer as a detector of the HPLC and/or the SEC systems: While dynamic light scattering method is widely utilized to characterize the polymeric materials and small particles in solution, the current static light scattering method is less significantly used for the characterization. This is likely due to the difficulty of the optical purification because the effect of dust particle may affect the serious error. The eluent of the SEC columns is optically purified since the large particles are adsorbed in the guard column or elute earlier than the objective polymer molecules. Thus, the SEC-MALS method is widely used to measure static light scattering of polymer samples in solution. This method is also applicable for the mixed solvent system because the eluted polymer solution is substantially the same as the dialyzed solution. The field-flow-fractionation (FFF) is also a powerful tool to purify the polymers and/or particles in dispersion media prior to measuring light scattering. Figure 3.9 shows the results from SEC-MALS for a polystyrene sample [12]. The obtained data is mostly fitted by the data from the conventional method [13–15].

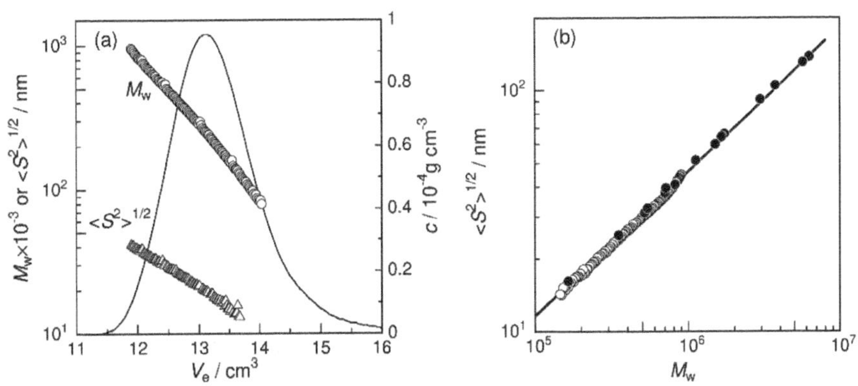

Fig. 3.9 **a** Plots of M_w, $\langle S^2\rangle^{1/2}$, and c versus V_e for a polystyrene sample in THF at 25 °C [12]. Molar mass dependence of $\langle S^2\rangle^{1/2}$ for polystyrenes in THF. Open circles: data for a polystyrene sample from SEC-MALS, filled circles: literature values from the conventional method [13–15]

3.3 Hydrodynamic Properties

3.3.1 Dynamic Light Scattering

The difference between static and dynamic light scattering is that the former measures the time average value of the scattered light intensity, while the latter observes the fluctuation of the scattered light intensity with time. The fluctuation of the scattered light is quantified by the following normalized time correlation function [16].

$$g^{(2)}(t, q) = \frac{\langle I(0, q)I(t, q)\rangle}{\langle I(t, q)\rangle^2} \quad (3.3.1)$$

In the equation, $I(t, q)$ is the scattered light intensity at time t and at the absolute value of the scattering vector, q.

Consider the scattered light from solute molecules in a solvent. When the solute molecules move according to the Fickian diffusion equation, $g^{(2)}(t)$ is expressed as follows [16].

$$g^{(2)}(t, q) = \exp(-2Dq^2t) + 1 \quad (3.3.2)$$

In the equation, D is the translational diffusion coefficient of the solute. According to this equation, by plotting $\ln[g^{(2)}(t, q) - 1]$ against q^2t, D can be obtained from the initial slope of the fitted straight line. An example of the plot [17] for a polystyrene (PS) polymacromonomer sample in cyclohexane (34.5 °C) is shown in Fig. 3.10. The data at different scattering angles could be fitted by straight lines with almost the same slope, from which D for this solute was determined. Strictly speaking, D at a finite solute mass concentration c is called the apparent diffusion coefficient D_{app}, and D is obtained from the D_{app} values extrapolated to $c \to 0$.

3.3.2 Sedimentation Velocity

The sedimentation coefficient can be measured using the analytical ultracentrifuge shown in Fig. 2.5. When measuring the sedimentation coefficient, the rotor is rotated at a higher rotational speed than in the case of the sedimentation equilibrium method described in Sect. 2.2.3. If the density of the solute is higher than that of the solvent, a steep sedimentation interface appears as shown in Fig. 3.11. The position of the sedimentation interface is expressed as the position r_{m} of the peak vertex of the concentration gradient dc/dr, which can be visualized with what is called the Schlieren optical system, where r is the distance from the center of rotation. If the system reaches the steady state, the sedimentation interface moves at a constant velocity toward the bottom of the cell over the centrifugal time t. The sedimentation velocity dr_{m}/dt is proportional to the acceleration of the centrifugal force $r\omega^2$ per unit mass, and the

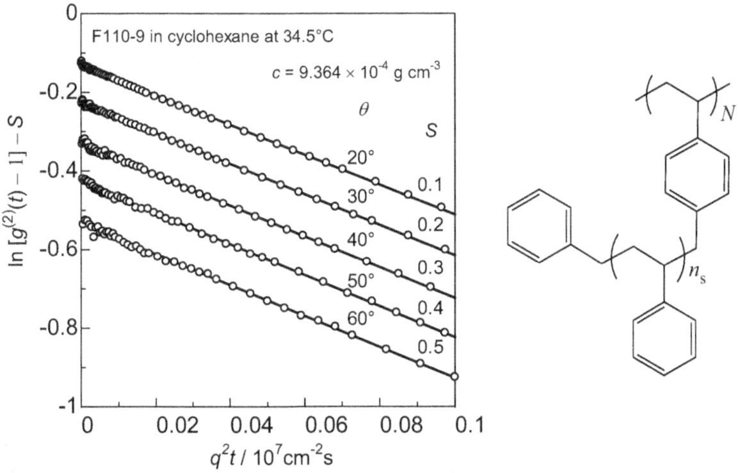

Fig. 3.10 Plots of $\ln[g^{(2)}(t, q) - 1]$ against q^2t for a polystyrene polymacromonomer sample in cyclohexane at 34.5 °C (left) and the chemical structure of the polymer (right) [17]

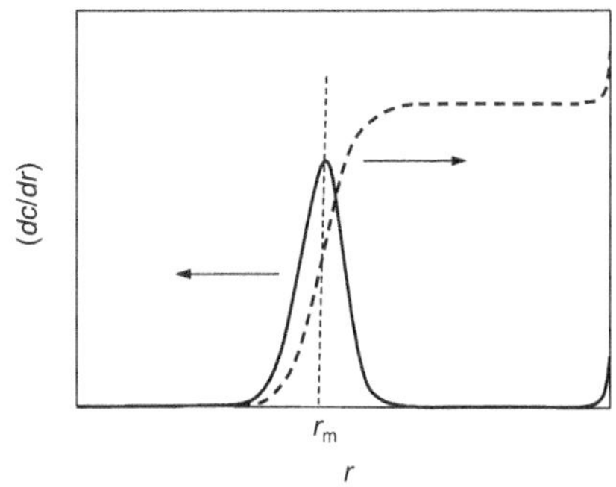

Fig. 3.11 Schematic representation of the concentration (dashed line) and the concentration gradient (dashed line) against the distance from the center of rotation for sedimentation velocity measurements

proportional coefficient is expressed as s, which is referred to as the sedimentation coefficient of the solute [18].

$$s = (dr/dt)/r\omega^2 \tag{3.3.3}$$

Here, ω is the angular velocity of the rotor. The above equation is rewritten as

$$\frac{d\ln r}{dt} = s\omega^2 \tag{3.3.4}$$

Therefore, s is obtained from the slope of the straight line fitted to the plot of ln r_{m} versus $\omega^2 t$. Figure 3.12 shows the relation between ln r_{m} versus t for a poly(hexyl isocyanate) (PHIC) sample in hexane [19] as an example. The experimental data are represented by the straight line, whose slope gives s.

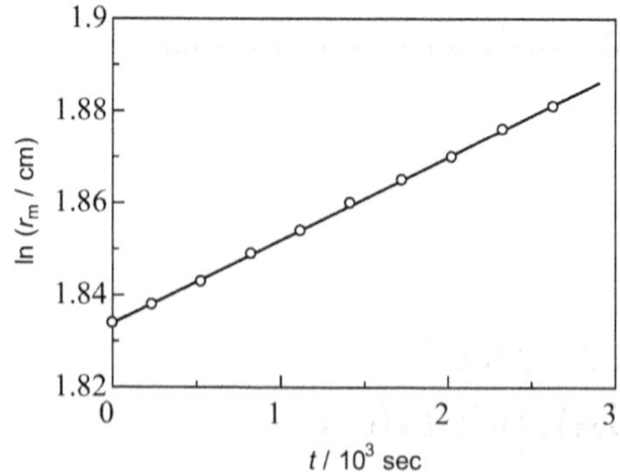

Fig. 3.12 Plot of ln r_m versus t for a PHIC sample in hexane [19]

The concentration dependence of s^{-1} is empirically expressed by the following equation:

$$s^{-1} = s_0^{-1}(1 + k_s c_0) \tag{3.3.5}$$

Here, c_0 means the initial solute concentration. The concentration dependence of s^{-1} against c_0 for PHIC with different molecular weight [19] is shown in Fig. 3.13. The extrapolated value to $c_0 = 0$ give $s_0{}^{-1}$ for each sample.

By the following equation, s_0 is related to the friction coefficient f:

$$s_0 = \frac{M(1 - \overline{v}\rho_0)}{N_A f} \tag{3.3.6}$$

Here, M and $\overline{v}$ are the molecular weight and the specific volume of the solute, respectively, ρ_0 is the density of the solvent, and $(1 - \overline{v}\rho_0)$ is the buoyancy factor; when this factor is positive, the solute molecules are sedimented to the bottom of the cell, on the other hand, when it is negative, they are buoyed to the meniscus of the cell.

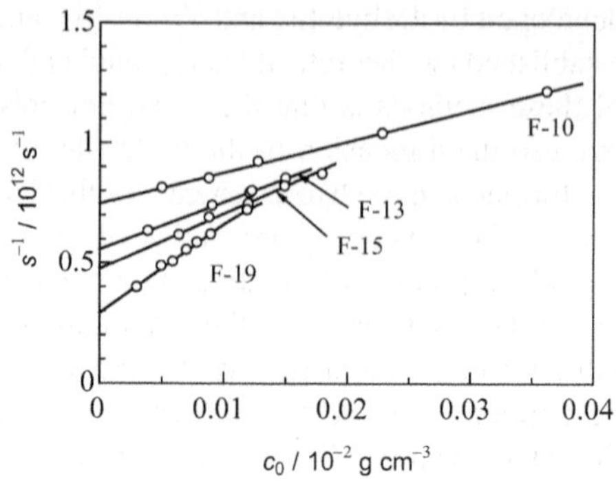

Fig. 3.13 Plots of s^{-1} versus c_0 for PHIC samples in hexane [19]

3.3.3 Hydrodynamic Radii

It is known that D for rigid particles is expressed by the following Einstein equation.

$$D = k_B T / f \tag{3.3.7}$$

The friction constant f for a rigid sphere with the radius a is given by the following Stokes equation:

$$f = 6\pi \eta_0 a \tag{3.3.8}$$

Here, η_0 is the viscosity of the solvent. Based on these equations, the hydrodynamic radius R_H is defined for any type of the solute molecule by the following equation.

$$R_H = k_B T / 6\pi \eta_0 D \tag{3.3.9}$$

From the sedimentation coefficient s_0, R_H is calculated by the following equation:

$$R_H = \frac{M(1 - \overline{v}\rho_0)}{6\pi \eta_0 N_A s_0} \tag{3.3.10}$$

Figure 3.14 shows the molecular weight dependence of R_H for PHIC in hexane [19] and PS in toluene [20] and in cyclohexane (Θ solvent) [21]. The polystyrene data can be fitted by the straight lines with slope 0.55 for the toluene solution and 0.5 for the cyclohexane solution. The data of the PHIC solution are fitted by a line with a larger slope.

R_H reflects the dimension of the solute molecule, although it does not agree with $\langle S^2 \rangle^{1/2}$. The ratio of the two quantities is denoted by ρ as

$$\rho = \langle S^2 \rangle^{1/2} / R_H \tag{3.3.11}$$

Fig. 3.14 Hydrodynamic raii for PHIC in hexane (triangles) [19] and PS in toluene (filled circles) [20] and cyclohexane (unfilled circles) [21]

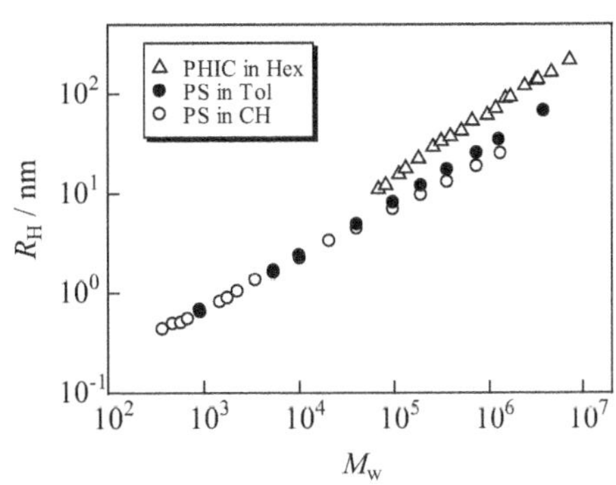

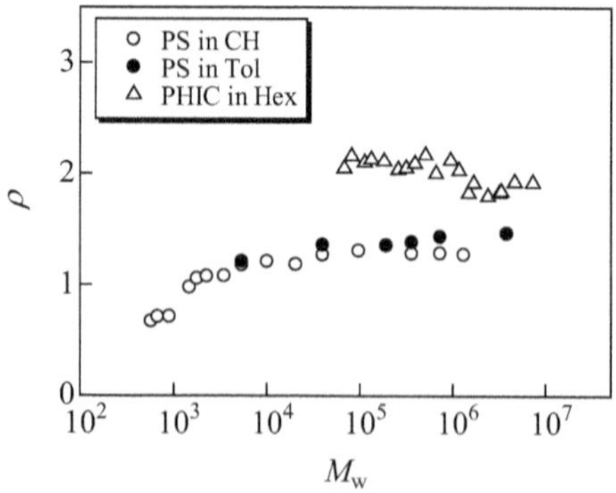

Fig. 3.15 ρ for PS in toluene [20] and cyclohexane [21] (filled and unfilled circles, respectively) and ρ for PHIC in hexane [19] plotted against M_w

Theoretically, the ρ value of spherical particles is known to be 0.775, and that of an infinitely long rod-like chain is ∞. For unperturbed random coil chains without the intramolecular excluded-volume effect, the theoretical value 1.505 of Kirkwood and Riseman (KR) is known [22].

Figure 3.15 shows the molecular weight dependence of ρ for several polymer-solvent systems. For PS in cyclohexane (CH), ρ values approach ca. 1.3 with increasing molecular weight, which is close to the theoretical value of KR. The ρ values for PS decrease with decreasing molecular weight because it approaches the rigid sphere value due to the effect of the chain thickness. On the other hand, for PHIC, which is a relatively stiff polymer, ρ decreases with increasing molecular weight and approaches the random coil value. This is because as the molecular weight decreases, the molecular shape approaches the rigid rod.

For rigid sphere particles, $[\eta]$ is described by the following Einstein's equation:

$$[\eta] = \frac{5}{2} v_\mathrm{e} \tag{3.3.12}$$

Here, v_e is the specific volume of the sphere and given with the radius of the sphere r_e by

$$v_\mathrm{e} = \frac{4\pi r_\mathrm{e}^3 N_\mathrm{A}}{3M} \tag{3.3.13}$$

From the above equations, the viscosity radius R_η can be defined for any macromolecules as

$$R_\eta = \left(\frac{3M}{10\pi N_\mathrm{A}} [\eta] \right)^{1/3} \tag{3.3.14}$$

However, $[\eta]$ is more often described by the following Flory-Fox equation.

Flory and Fox proposed to the following relation between $[\eta]$ and $\langle S^2 \rangle$; [23] by the analogy of Eq. 3.3.12.

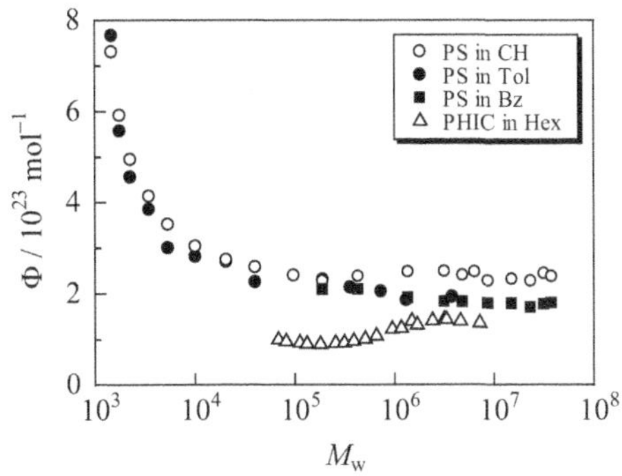

Fig. 3.16 Φ for polystyrene in good solvents (toluene [24, 25] and benzene [26, 27] indicated by filled circles and squares, respectively) and in the theta solvent (cyclohexane [26–29] indicated by unfilled circles) and for poly(hexyl isocyanate) in hexane (triangles) [19] plotted against M_w

$$[\eta] = \Phi \frac{(6\langle S^2 \rangle)^{3/2}}{M} \tag{3.3.15}$$

Here, Φ is called the Flory-Fox factor. Figure 3.16 shows Φ for PS in good (toluene and benzene) and (cyclohexane) theta solvents and that for PHIC in hexane plotted against M_w. It is seen that Φ for PS approaches 2.5×10^{23} mol^{-1} in good solvents and about 1.8×10^{23} mol^{-1} in the theta solvent at high M_w. With decreasing molecular weight, PS values increase approaching the rigid sphere value, 9.23, because the contour length of the polymer approaches the polymer diameter. On the other hand, values for PHIC once decrease with decreasing molecular weight approaching the rigid rod value, 0 and increase as is the case for PS after passing the minimum.

3.4 Determination of Molecular Parameters

3.4.1 Wormlike Chain Parameters

3.4.1.1 Molecular Weight Dependence of Radius of Gyration

It is known that semi-flexible polymers, which have intermediate properties between flexible polymers and rod-like polymers, are modeled by the wormlike chain or Kratky-Porod (KP) chain model [30]. As mentioned in 3.1, the wormlike chain model can be characterized by the stiffness parameter of the chain, λ^{-1}, and the molecular weight per unit contour length, M_L, that relates the contour length of the chain, L, to the molecular weight, M by $L = M/M_L$.

The mean-square radius of gyration of wormlike chains is expressed by the following Benoit-Doty equation [31].

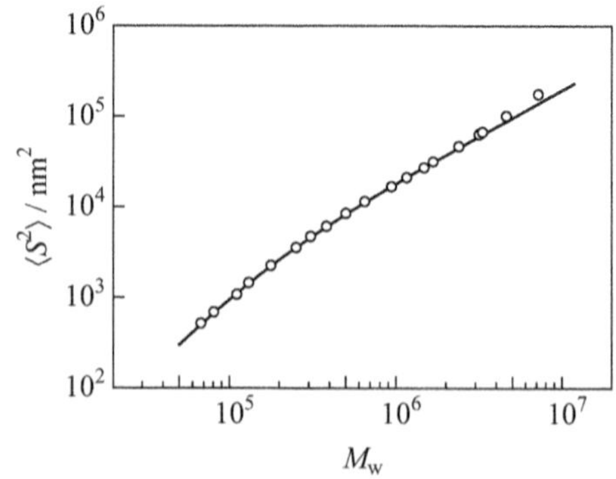

Fig. 3.17 Mean-square radius of gyration for PHIC in hexane [19]. Solid line represents the value calculated by Eq. 3.4.1 with the parameters $\lambda^{-1} = 84$ nm and $M_L = 715$ nm^{-1}

$$\langle S^2 \rangle = \frac{L}{6\lambda} - \frac{1}{4\lambda^2} + \frac{1}{4\lambda^3 L} - \frac{1}{8\lambda^4 L^2}\left[1 - \exp(-2\lambda L)\right] \tag{3.4.1}$$

Figure 3.17 shows $\langle S^2 \rangle$ data for a typical semiflexible polymer PHIC in hexane [19]. The solid line in the figure shows the calculated value, which almost quantitatively represents the experimental values indicated by the circles. The deviation of the calculated value for $M_w > 3 \times 10^6$ from the observed values is due to the excluded-volume effects.

Flexible polymer chains in good solvents are expanded by the repulsive intramolecular interaction among polymer segments. Such effect, called the excluded-volume effect, can be taken account by multiplying $\langle S^2 \rangle$ by the square of the radius expansion factor α_s as

$$\langle S^2 \rangle = \alpha_S^2 \langle S^2 \rangle_0 \tag{3.4.2}$$

Here, the subscript 0 means the value without the excluded-volume effect, or the unperturbed value, which is given by Eq. 3.4.1.

The radius expansion factor for flexible chains is known to be accurately expressed by the following Domb-Barrett equation [32]:

$$\begin{aligned}\alpha_S^2 = &\left[1 + 10z + \left(\frac{70\pi}{9} + \frac{10}{3}\right)z^2 + 8\pi^{3/2}z^3\right]^{2/15} \\ &\times\left[0.933 + 0.067\exp(-0.85z - 1.39z^2)\right]\end{aligned} \tag{3.4.3}$$

In the above equation, z is the excluded-volume parameter defined by

$$z = \left(\frac{3}{2\pi}\right)^{3/2}(\lambda B)(\lambda L)^{1/2} \tag{3.4.4}$$

Here, B is the excluded-volume strength, which is defined for the wormlike chain with the binary-cluster integral β and the bead spacing a, by

$$B = \beta/a^2 \tag{3.4.5}$$

For wormlike chains, z in Eq. 3.4.3 is replaced by the following scaled excluded-volume parameter [4]:

$$\tilde{z} = \frac{K(\lambda L)}{(4/3)} z \tag{3.4.6}$$

Here, $K(\lambda L)$ and (4/3) are the first-order perturbation coefficients for the wormlike chain and the flexible chains, respectively. The former coefficient is given by the following equation as a function of the reduced contour length (λL) [4, 33]:

$$K(\lambda L) = \begin{cases} \dfrac{4}{3} - \dfrac{2.711}{(\lambda L)^{1/2}} + \dfrac{7}{6\lambda L} & (\lambda L > 6) \\ \dfrac{1}{(\lambda L)^{1/2}} \exp\left[-\dfrac{6.611}{\lambda L} + 0.9198 + 0.03516(\lambda L) \right] & (\lambda L \le 6) \end{cases} \tag{3.4.7}$$

Polystyrene polymacromonomer (PSPM) molecules (the chemical structure is given in Fig. 3.10) behave as wormlike chains in solution. Figure 3.18 shows the double-logarithmical plots of $\langle S^2 \rangle$ for PSPMs having different side chain length against the weight-average degree of polymerization of the main chain, N_w [17, 34–36], where F15, F33, F65, and F110 denote PSPM with 15, 33, 65, and 110 (weight-average values) monomeric units in a side chain. The solid lines in the figure show the calculated values by Eq. 3.4.1, which fit the experimental values except the data points for toluene solutions of F15 and F33 and the toluene and cyclohexane solutions of F110 with N_w smaller than 500. The data for toluene solutions of F15 and F33 are satisfactorily fitted by the dot-dashed lines considering the excluded-volume effect by Eq. 3.4.2 with 3.4.3. The data for F110 can be fitted by the dashed lines, which were calculated by considering the effect of the chain thickness according to the following equation [29]:

$$\langle S^2 \rangle = \langle S^2 \rangle_c + d^2/8 \tag{3.4.8}$$

Here $\langle S^2 \rangle_c$ is the mean-square radius of the chain contour given by Eq. 3.4.1 and d means the chain diameter. The molecular parameters used for fitting the data are summarized in Table 3.5.

3.4.1.2 Molecular Weight Dependence of Intrinsic Viscosity

The intrinsic viscosity of Kratky-Porod (KP) wormlike cylinders is formulated by Yamakawa, Fujii, and Yoshizaki [4, 37, 38] as

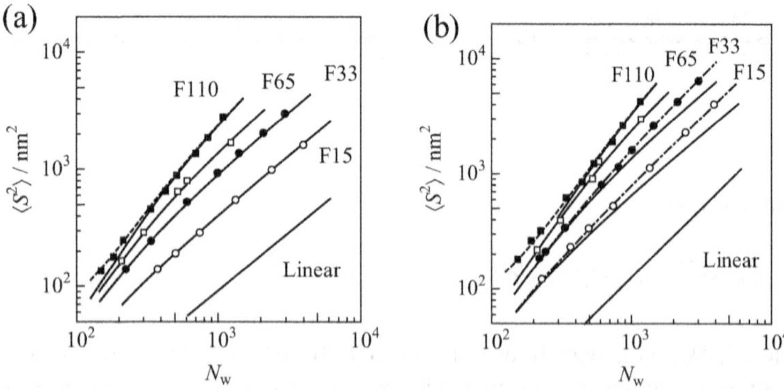

Fig. 3.18 Mean-square radius of gyration for polystyrene polymacromonomers with different number of monomeric units on a side chain, n_s (15, 33, 65, and 113 for F15, F33, F65, and F110, respectively), in cyclohexane at 34.5 °C (**a**) and in toluene at 15 °C (**b**) [17, 34–36] plotted double-logarithmically against the weight-average degree of polymerization of main chain

Table 3.5 Wormlike chain parameters determined from $\langle S^2 \rangle$ for polystyrene polymacromonomers in cyclohexane at 34.5 °C (Θ solvent) and toluene at 15 °C (good solvent)

	Cyclohexane at 34.5 °C		Toluene at 15 °C	
	M_L/nm^{-1}	λ^{-1}/nm	λ^{-1}/nm	B/nm
F15	6,200	9.5	16	4.5
F33	13,000	22	36	18
F65	25,000	36	75	–
F110	45,500	80	155	–

$$[\eta]_{\mathrm{KP}} = \frac{\Phi(\lambda L)^{3/2}}{M}\left(1 - \sum_{j=1}^{4} C_j L^{-j/2}\right)^{-1} \quad (\lambda L \geq 2.278) \tag{3.4.9}$$

where $\Phi = 2.87 \times 10^{23}$ mol^{-1}(Flory's viscosity factor [23] in the coil limit) and

$$C_j = \sum_{k=0}^{2} \alpha_{jk}(\lambda d)^k + \sum_{k=0}^{1} \beta_{jk}(\lambda d)^{2k} \ln(\lambda d) \quad (j = 1 - 4) \tag{3.4.10}$$

The coefficients in the above equation are given in Appendix 3.A. The above equations are applicable for $\lambda L \geq 2.278$ when $\lambda d < 0.2$ and $(\lambda L)^{1/2} \gtrsim 30\lambda d$ when $0.2 < \lambda d < 1.0$. The equation for $\lambda L < 2.278$ can be found in Ref. [4] or [38].

Figure 3.19 shows an example for the analysis of the intrinsic viscosity of poly(*p*-phenylene benzoxiazole) (PBO) [24]. By fitting the calculated line to the experimental data λ^{-1}, d, and M_L were determined.

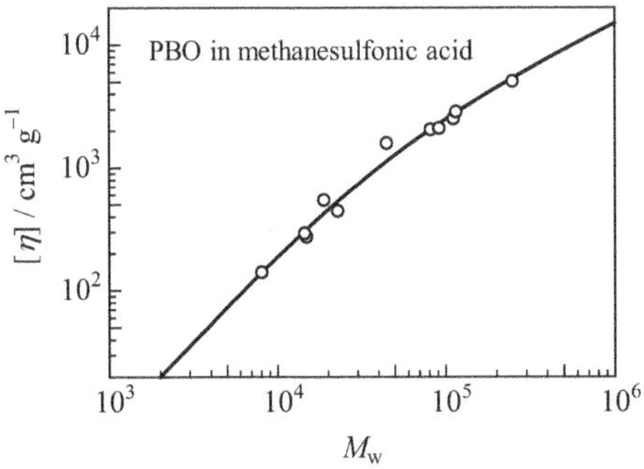

Fig. 3.19 Intrinsic viscosity of PBO in methanesulfonic acid double-logarithmically plotted against the weight-average molecular weight. Solid line represents the value calculated by Eq. 3.4.9 with the parameters $\lambda^{-1} = 40$ nm, $d = 0.6$ nm, and $M_L = 203$ nm^{-1} [39]

The applicability of the above equations is limited to relatively short chains with d much smaller than λ^{-1}. This excludes very flexible polymers and very thick stiff chains. The theory applicable for such cases is limited to that for the touched-bead model formulated by Yoshizaki et al. [40]. In this theory, $[\eta]$ is written as

$$[\eta] = [\eta]_{KR} + (L/d)[\eta]_E \quad (\lambda d > 0.2) \tag{3.4.11}$$

Here, $[\eta]_{KR}$ is the solution of the Kirkwood-Riseman equation [22] for the Kratky-Porod (KP) chain given by

$$[\eta]_{KR} = \frac{6^{3/2}\Phi\langle S^2\rangle_{KP}^{3/2}}{M}\Gamma(\lambda L, \lambda d) \tag{3.4.12}$$

where $\langle S^2\rangle_{KP}$ is $\langle S^2\rangle$ calculated by Eq. 3.4.1 and Γ is a function of λL and λd, expressed as

$$\Gamma(\lambda L, \lambda d)^{-1} = 1 + e^{-5\lambda L}\sum_{i=0}^{3} c_i(\lambda L)^{i/2} + e^{-1/4\lambda L}\sum_{i=4}^{7} c_i(\lambda L)^{-(i-3)/2} \tag{3.4.13}$$

With

$$c_i = \sum_{j=0}^{2} \alpha_{ij}(\lambda d)^j + \sum_{j=0}^{1} \beta_{ij}(\lambda d)^{2j}\ln(\lambda d) \tag{3.4.14}$$

The coefficients in Eq. 3.4.14 is given in Appendix 3.B.

When the excluded-volume effect cannot be ignored, $[\eta]_0$is multiplied by the intrinsic viscosity expansion factor $\alpha\eta^3$ as

$$[\eta] = \alpha_\eta^3[\eta]_0 \tag{3.4.15}$$

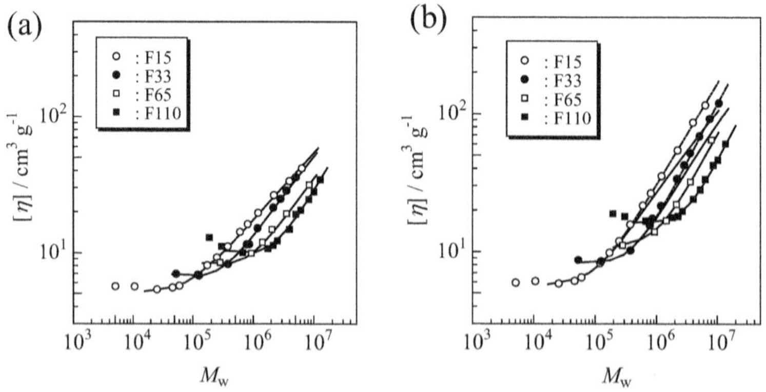

Fig. 3.20 Intrinsic viscosity for polystyrene polymacromonomers with different number of monomeric units in a side chain n_s (15, 33, 65, and 113 for F15, F33, F65, and F110, respectively) in cyclohexane at 34.5 °C a and in toluene at 15 °Cb [17, 36, 42] plotted double-logarithmically against the weight-average degree of polymerization of main chain

To calculate α_η, the following Barrett's equation [41] can be used.

$$\alpha_\eta^3 = (1 + 3.8\tilde{z} + 1.9\tilde{z}^2)^{0.3} \tag{3.4.16}$$

Here, $\tilde{z}$ is given by Eq. 3.4.6.

Figure 3.20a, b shows the intrinsic viscosity of polystyrene polymacromonomers in cyclohexane and toluene, respectively. The solid lines in (a) indicate the calculated values by Eqs. 3.4.11–3.4.14, which closely fit the data points. The contour length of the polymacromonomer chains is calculated by $L = M/M_L + \delta$, where δ is the extended contour length caused by the side chains covering ends of the polymacromonomer molecule. For each polymacromonomer, the molecular weight dependence of $[\eta]$ at low M_w is very weak, because the molecular shape approaches the sphere with decreasing molecular weight. At higher M_w, $[\eta]$ increases according to the Mark-Houwink-Sakurada equation, Eq. 2.2.28. For toluene solutions of F15 and F33, the excluded-volume effect is taken into account by Eq. 3.4.15 with Eq. 3.4.16. The calculated values indicated by the dot-dashed lines closely fit the data points. The fitting parameters are given in Table 3.6.

3.4.1.3 Molecular Weight Dependence of Hydrodynamic Radius

From the translational friction coefficient of wormlike cylinders formulated by Yamakawa and Fujii [43], the hydrodynamic radius R_H is expressed by the following equation:

Table 3.6 Wormlike chain parameters determined from $[\eta]$ for polystyrene polymacromonomers in toluene at 15 °C (good solvent) and cyclohexane at 34.5 °C (Θ solvent)

	Cyclohexane at 34.5 °C		Toluene at 15 °C	
	d/nm	δ/nm	d/nm	δ/nm
F15	4.7	2.2	5.0	2.3
F33	7.4	4.0	8.5	4.3
F65	11.5	6.0	14.0	6.0
F110	15	13	21	14

$$\begin{aligned} R_{\mathrm{H}}^{-1} &= \frac{2}{L}\left[C_1 \ln\sigma + C_2 + C_3(\lambda L) + C_4(\lambda L)^2 + C_5(\lambda L)^3 + C_6\sigma^{-1}\ln\sigma \right. \\ &\quad \left. + C_7\sigma^{-1} + C_8\sigma^{-2} + C_9\sigma^{-3} + C_{10}\sigma^{-4}\right] \qquad \text{for } \lambda L \le 2.278 \\ &= \frac{2}{L}\left[A_1(\lambda L)^{1/2} + A_2 + A_3(\lambda L)^{-1/2} + A_4(\lambda L)^{-1} + A_5(\lambda L)^{-3/2}\right] \\ &\qquad \text{for } \lambda L > 2.278 \end{aligned} \quad . \tag{3.4.17}$$

where $\sigma = L/d$. The coefficients in the above equations are functions of λd and given in Appendix 3.C.

Since the above theory does not consider the effect from chain ends, it cannot be used for short cylinders with L/d less than 4. For such short cylinders, the theory of Norisuye et al. [44] for the translational friction coefficient of wormlike cylinders capped with hemispheres at both ends is available, which expresses R_{H} as

$$R_{\mathrm{H}}^{-1} = \frac{2}{L}\left[C_1' \ln\sigma + C_2' + C_3'(\lambda L) + C_4'(\lambda L)^2 + C_5'(\lambda L)^3 + C_6'(\lambda L)^4 + C_7'(\lambda L)^5\right] \tag{3.4.18}$$

The coefficients in the above equation are functions of σ and λL and given in Appendix 3.D.

For the cases that the excluded-volume effect is not negligible, R_{H} is expressed as

$$R_{\mathrm{H}} = (R_{\mathrm{H}})_0\alpha_{\mathrm{H}} \tag{3.4.19}$$

Here, $(R_{\mathrm{H}})_0$ means R_{H} in the unperturbed state (given by Eq. 3.4.17 or 3.4.18) where no excluded-volume effect works. The hydrodynamic radius expansion factor α_{H} is expressed by the following Barrett's equation [41] with the modified excluded-volume parameter $\tilde{z}$ given by Eq. 3.4.6.

$$\alpha_{\mathrm{H}} = (1 + 6.02\tilde{z} + 3.59\tilde{z}^2)^{0.1} \tag{3.4.20}$$

Figure 3.21 shows R_{H} values for polystyrene polymacromonomers plotted double-logarithmically against the weight-average degree of polymerization of the main chain. The solid and dashed lines in the figure represent the calculated values of

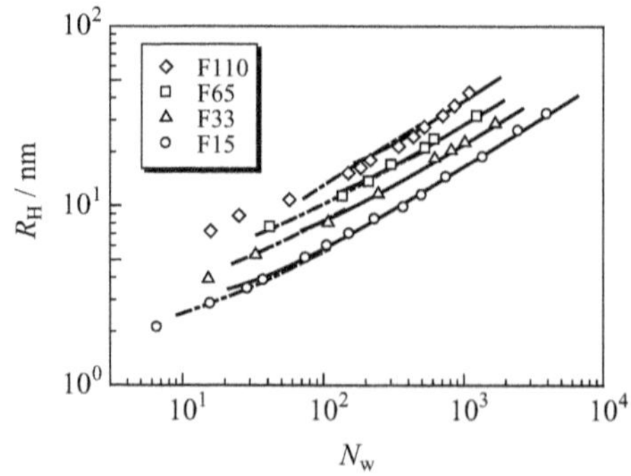

Fig. 3.21 Hydrodynamic radius for polystyrene polymacromonomers in cyclohexane at 34.5 °C plotted double-logarithmically against the weight-average degree of polymerization of main chain [17, 45, 46]. The values of the degree of polymerization of the side chain are as follows: 113 for F110; 65 for F65; 33 for F33; 15 for F15. The solid and dashed lines show the theoretical value of Yamakawa-Fujii [43] and Motowoka-Norisuye-Fujita [44], respectively. The fitting parameters are as follows: $\lambda^{-1} = 80$ nm, $d = 24$ nm for F110; $\lambda^{-1} = 36$ nm, $d = 11.5$ nm for F65; $\lambda^{-1} = 22$ nm, $d = 7.4$ nm for F33; and $\lambda^{-1} = 9.6$ nm, $d = 4.7$ nm for F15

Eqs. 3.4.17 and 3.4.18, respectively, showing that the former equation is valid for large N_w and the latter is valid for small N_w.

3.4.2 Analysis of Long Chain Branching

If compared at the same molecular weight, polymer molecules with branching structures have smaller sizes than linear polymers. Therefore, as a measure of the degree of branching, the ratio g_S of $\langle S^2 \rangle$ for branched polymer, $\langle S^2 \rangle_{\text{branched}}$, to that for linear polymer consisting of the same monomers, $\langle S^2 \rangle_{\text{linear}}$, is used. With increasing the branching degree, g_S becomes smaller. If branching occurs with equal probability at any monomeric unit, g_S for such a randomly branched polymer can be expressed as [47]

$$g_S = \frac{6}{m}\left\{\frac{1}{2}\left(\frac{2+m}{m}\right)^{1/2} \ln\left[\frac{(2+m)^{1/2}+m^{1/2}}{(2+m)^{1/2}-m^{1/2}}\right] - 1\right\} \qquad (\text{random}, f = 3) \tag{3.4.21}$$

where m is the number of branching points in a molecule. In the above equation, the branching point is assumed to consist of three branches, i.e., the functionality f of the branching unit is 3. When $f = 4$, it is expressed as

$$g_S = \frac{1}{m}\ln(1+m) \qquad (\text{random}, f = 4) \tag{3.4.22}$$

When the polymer molecule has a comb-like structure, g_S is written as [48]

$$g_S = [m(3m-2)r^3 + m(m+1)(m+2)r^2 + m(m+1)(2m+1)r + (m+1)^3] \\ /(rm+m+1)^3 \qquad \text{(regular comb)} \tag{3.4.23}$$

where r is the ratio of the number of monomers, n_s, in a side chain to the number of monomers, n_c, in the partial chain between the adjacent branching points of the main chain, i.e., $r = n_s/n_c$. Here, n_s and n_c are assumed to be constants over the molecule. Such molecular structure is called a regular comb. When the branching points are randomly placed on the main chain, g_S is given by

$$g_S = [3m - 2 + (m-6)(m-1)\phi + (5m-6)\phi^2 - (m-2)\phi^3]/m^2 \qquad \text{(random comb)} \tag{3.4.24}$$

where ϕ is the number of main-chain monomers divided by the total monomers in a molecule.

Figure 3.22 shows $\langle S^2\rangle^{1/2}$ for polyethylene (PE) samples synthesized under different pressures measured by SEC-MALS [49] double-logarithmically plotted against M_w. The solid line in the figure indicates the value for linear PE [50]. g_S values for each polymer sample are calculated with this line and are shown in Fig. 3.23.

The decrease of g_s with decreasing the polymerization pressure indicates an increase in the branching degree. The dashed lines in Fig. 3.23 show the calculated values by Eq. 3.4.21 assuming that m divided by the total number of monomers n in a molecule for each sample does not depend on molecular weight. However, the calculated values do not explain the observed data. The solid lines in the figure show the calculated values by Eq. 3.4.23 assuming that m/n is 0.1 for all polymers with the r values given in the figure legend. The close fits of the calculated values to the experimental data strongly suggest that these polymers have comb-like structures. It

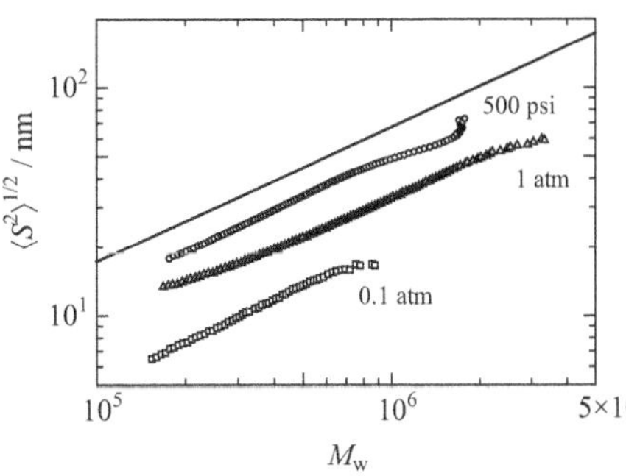

Fig. 3.22 $\langle S^2\rangle^{1/2}$ for polyethylene polymerized at the indicated pressures logarithmically plotted against M_w [49]. The solid line shows the value for linear polyethylene [50]

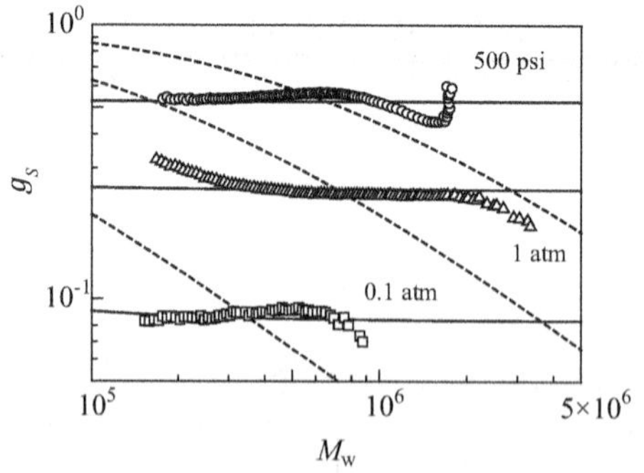

Fig. 3.23 g_S calculated from the data in Fig. 3.22 for polyethylene samples polymerized at the indicated pressures double-logarithmically plotted against M_w. Dashed lines indicate the calculated values by Eq. 3.4.21 with $m/n = 0.00025$, 0.01, and 0.1 from top to bottom and solid lines show the calculated values by Eq. 3.4.23 with $r = 0.9$, 3, and 11 from top to bottom and $m/n = 0.1$

is known, in the polymerization of ethylene with palladium catalyst, that β-hydrogen elimination occurs. By this process, macromonomer-like structure is made at the chain end. When the main chain is generated, macromonomers are also polymerized along with ethylene monomers. This may be the reason that these polymers have comb-like structures.

Using online viscometer and a MALS detector connected to SEC, the relation between $[\eta]$ and M_w can be obtained by a single injection. Figure 3.24 shows $[\eta]$ for the same PE samples as those in Fig. 3.22 [49] along with the value for linear PE indicated by the solid line [50]. From the $[\eta]$ values for branched polymers and linear one, we can obtain a similar property g_η, which is defined by $[\eta]$ for the branched polymer divided by that for the linear polymer consisting of the same numbers of monomers. From these data, g_η for these polymers are calculated and shown in Fig. 3.25.

It is known that the relation between g_η and g_s is expressed as

$$g_\eta = g_s^\epsilon \tag{3.4.25}$$

Fig. 3.24 $[\eta]$ data for the same samples as those in Fig. 3.22 [49]. The solid line shows the value for linear polyethylene [50]

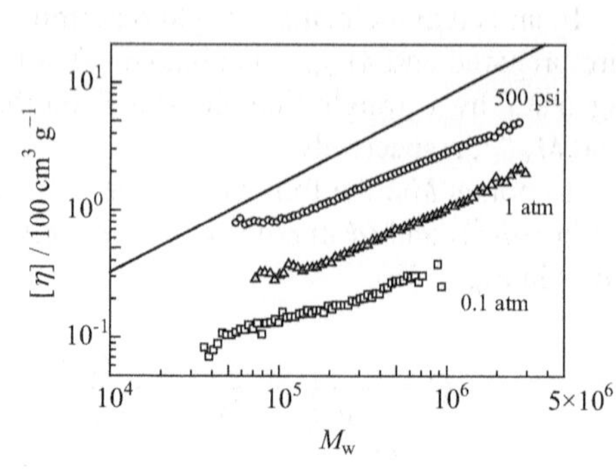

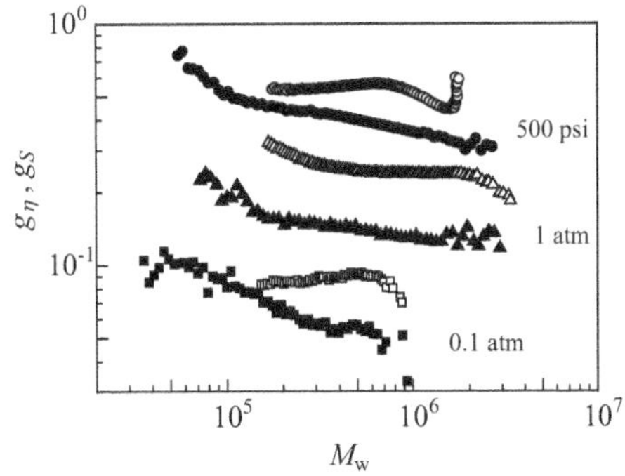

Fig. 3.25 g_η calculated from the data in Fig. 3.24 for polyethylene samples polymerized at the indicated pressures double-logarithmically plotted against M_w (filled symbols). Unfilled symbols show g_S.

The Flory-Fox relation (Eq. 2.2.30) predicts that ε is 3/2 while the Zimm-Kilb theory [51] for star polymers predicts $\varepsilon = 1/2$. It is known that ε for branched polymers having sphere-like structures is close to 1/2 while ε for branched polymers with thin long structure like comb polymers is close to 3/2.

In Fig. 3.26a, log $g\eta$ for comb polystyrenes (including PS-PM) and star polymers with various arm numbers are plotted against log g_s. The figure shows that the data for comb polymers are close to the line with the slope 3/2. On the other hand, values for star polymers are close to the line with the slope 1/2 when g_s is close to 1 (for small arm numbers) and approach the line with slope 3/2 with decreasing g_s (for large arm numbers). In Fig. 3.26b, similar plots for the PE samples in the above study are shown. The data for these samples are close to the line with the slope 3/2, which also suggests that the molecular structures of these PE samples are comb-like.

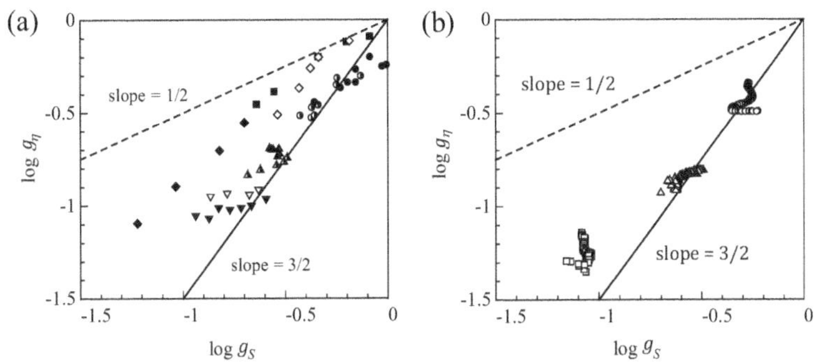

Fig. 3.26 Plots of log g_η against log g_s for comb polystyrene (half-filled [52] and filled circles [53]), PS-PM (filled triangles, F15; half-filled triangles, F33, unfilled down triangles, F64, filled down triangles, F110) [17, 34–36, 42], and star polymers (filled squares, PS; unfilled diamonds, PIP; filled diamonds, PB) [54] in theta solvents (**a**). The same type of plots for the polyethylene samples (the symbols are the same as those in Figs. 3.22 and 3.24) (**b**). The solid and dashed lines have the slopes 3/2 and 1/2, respectively

3.4.3 Analysis of Aggregates

In dynamic light scattering, the observed $g^{(2)}(t, q)$ can be expressed as a linear combination of exponentially decaying functions $\exp(-t/\tau_i)$ as

$$[g^{(2)}(k,t)-1]^{1/2}=\sum_i A(\tau_i)e^{-t/\tau_i} \tag{3.4.26}$$

Each term corresponds to the contribution from a solute component with a different diffusion constant and the coefficient $A(\tau_i)$ corresponds to the scattering intensity from that component. The decomposition is made by a specialized computer program, such as CONTIN [55]. Since the scattering intensity is proportional to the mass concentration multiplied by the molecular weight, this method is suitable for the analysis of large aggregates.

Figure 3.27 shows the data for poly(hexyl isocyanate) (PHIC) in hexane at 15 °C and 155 h after quenching the solution from 30 °C [56]. In (a), $g^{(2)}(t, q)$ at the scattering angle $\theta = 40°$ decays in two steps. The decomposition result is shown in (b), where the left and right peaks show the fast and slow components, respectively. The $q^2\tau$ value at the peak top of $A(\tau)$ is equated to D^{-1}, which is used to calculate R_H for each component by Eq. 3.3.9.

Figure 3.28 shows the results for static scattering measurement for the same solution at the different quenching time t_{quench} [56]. Here, $\hat{S}(k)$ is the quantity defined by

$$\hat{S}(k)=R_\theta/(KcM_0) \tag{3.4.27}$$

where M_0 denotes the molecular weight of the PHIC monomer. Kanao et al. proposed to divide $\hat{S}(k)$ into the fast component $\hat{S}_{fast}(k)$ and slow components $\hat{S}_{slow}(k)$ using the peak areas in Fig. 3.27b. The results are shown in Fig. 3.29.

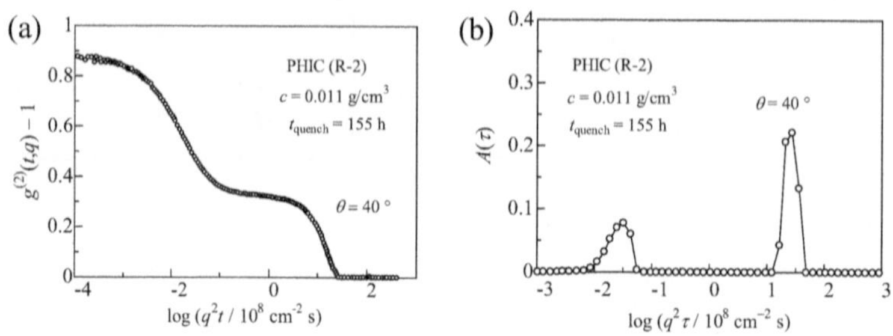

Fig. 3.27 Dynamic light scattering data for PHIC ($M_w = 23.6 \times 10^4$) in hexane (15 °C) [56]. The panels **a** and **b** show plots of $g^{(2)}(t, q)$ versus q^2t and $A(\tau)$ versus $q^2\tau$, respectively. The left and right peaks in panel (**b**) correspond to the fast and the slow components, respectively

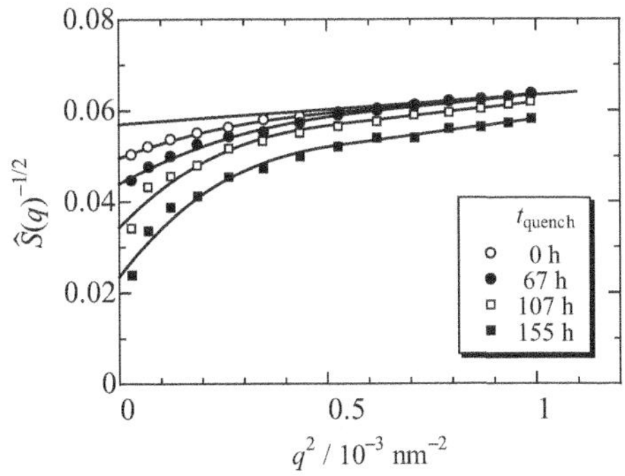

Fig. 3.28 Static light scattering data for the same solution as that in Fig. 3.26 at the different quenching time t_{quench} [56]

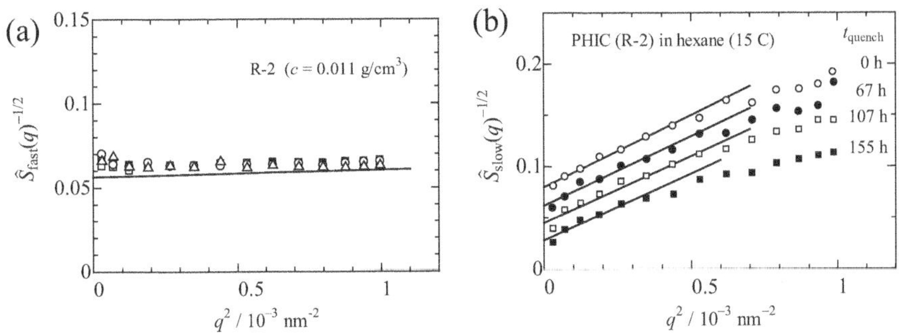

Fig. 3.29 The reciprocal scattering intensities for the same solution as that in Fig. 3.26, decomposed into the fast (**a**) and slow (**b**) components [56]

As for the fast component, $\hat{S}_{fast}(k)^{-1/2}$ is almost independent of t_{quench}. The solid line in (a) represents the expected value for the isolated molecule. The close agreement between the line and the experimental values shows that the fast component corresponds to the isolated molecules. On the other hand, $\hat{S}_{slow}(k)^{-1/2}$ in (b) significantly changes with t_{quench}. The initial slope of the fitted lines (solid lines) can be used to determine $\langle S^2 \rangle$ for these slow components or aggregates. The obtained $\langle S^2 \rangle$ values are used to calculate ρ for the aggregates using R_H values determined by DLS. Figure 3.30 shows ρ for samples with different molecular weight as functions of t_{quench}. ρ values for R-2 ($M_w = 23.6 \times 10^4$) are almost independent of t_{quench} and around 3, which suggests that the aggregate shape is close to rod. The values for O-2 ($M_w = 4.5 \times 10^4$) are around 0.5 showing that the aggregate shape is close to sphere. The values for JP-3-3 ($M_w = 6.3 \times 10^4$) are initially close to 1.2 and decrease with t_{quench}, showing that the aggregate shape changes from random-coil like to sphere like with the time.

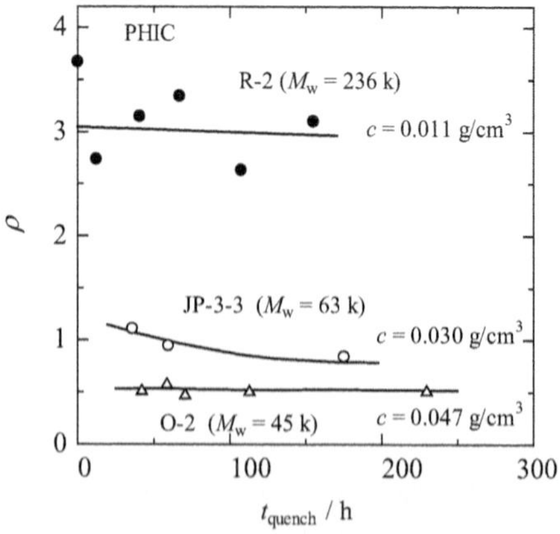

Fig. 3.30 ρ as functions of t_{quench} for PHIC samples with different molecular weights [56]

Appendix

3.A Coefficients in Eq. 3.4.10

Values for $\lambda d \leq 0.1$

j	α_{j0}	α_{j1}	α_{j2}	β_{j0}	β_{j1}
1	3.230981	−143.7458	−1906.263	2.463404	−1422.067
2	−22.46149	1347.079	19387.400	−5.318869	13868.57
3	54.81690	−3235.401	−49357.06	15.41744	−34447.63
4	−32.91952	2306.793	36732.64	−8.516339	25198.11

3.B Coefficients in Eq. 3.4.14

i	α_{i0}	α_{i1}	α_{i2}	β_{i0}	β_{i1}
0	−9.6291	1.6198 (2)	1.1316 (2)	−1.5358	9.4913 (2)
1	−2.3491	−1.4420 (2)	−2.0502 (3)	−2.3605	−3.4732 (3)
2	5.4811 (1)	−4.8402 (2)	4.1942 (3)	1.0550 (1)	4.0771 (3)
3	−6.2255 (1)	7.8877 (2)	−2.6846 (3)	−1.1528 (1)	−1.1290 (3)
4	3.0814 (−1)	−4.5617	1.5182	−1.9421	−3.1301
5	−5.1619	1.6758 (1)	−4.0308	5.1951 (−1)	1.2811 (1)
6	2.9298	−1.3380 (1)	−2.6757	1.1938 (−1)	−9.9978
7	−6.2856 (−1)	1.6070	7.4332	−8.2021 (−2)	−2.8832

a (n) stands for $a \times 10^n$

3.C Coefficients in Eq. 3.4.17 as Functions of λd

$C_1 = 1 - 0.01412(\lambda d)^2 + 0.00592(\lambda d)^4$
$C_2 = 0.3863 - 0.1667(\lambda d) + 0.0016(\lambda d)^2 - 0.0224(\lambda d)^3 - 0.0007(\lambda d)^4$
$C_3 = 0.1667 + 0.0222(\lambda d)^2 + 0.0017(\lambda d)^4$
$C_4 = 0.01883 - 0.00789(\lambda d)^2 - 0.00038(\lambda d)^4$
$C_5 = -0.002039 + 0.000805(\lambda d)^2 + 0.000017(\lambda d)^4$
$C_6 = 0.04167(\lambda d) + 0.00567(\lambda d)^3$
$C_7 = 0.5 + 0.0786(\lambda d) - 0.0094(\lambda d)^2 + 0.0107(\lambda d)^3 + 0.0039(\lambda d)^4$
$C_8 = -0.0625 + 0.00132(\lambda d)^2 - 0.00055(\lambda d)^4$
$C_9 = 0.001302(\lambda d) + 0.00018(\lambda d)^3$
$C_{10} = 0.001953 - 0.000064(\lambda d)^2 + 0.000027(\lambda d)^4$
$A_1 = 1.843$
$A_2 = -[1 - 0.01412(\lambda d)^2 + 0.00592(\lambda d)^4] \ln(\lambda d) - 1.0561 - 0.1667(\lambda d) - 0.1900(\lambda d)^2 - 0.0224(\lambda d)^3 + 0.0190(\lambda d)^4$
$A_3 = 0.1382 + 0.6910(\lambda d)^2$
$A_4 = -[0.04167(\lambda d)^2 + 0.00567(\lambda d)^4] \ln(\lambda d) - 0.3301 + 0.5(\lambda d) - 0.5854(\lambda d)^2 - 0.0094(\lambda d)^3 - 0.0421(\lambda d)^4$
$A_5 = -0.0300 + 0.1209(\lambda d)^2 + 0.0259(\lambda d)^4$

3.D Coefficients in Eq. 3.4.18 as Functions of σ (= L/d) and λd

$C_1' = 1 - 0.1250\sigma^{-1}(\lambda d) + 0.1406\,\sigma^{-1}(\lambda d)^3$
$C_2' = 0.3863 + 0.6863\,\sigma^{-1} - 0.06250\,\sigma^{-2} - 0.01042\,\sigma^{-3} - 0.0006510\,\sigma^{-4} + 0.0005859\,\sigma^{-5}$
$C_3' = 0.1667 - 0.06838\,\sigma^{-2} + 0.02083\,\sigma^{-3} - 0.01693\,\sigma^{-4} - 0.008594\,\sigma^{-5}$
$C_4' = 0.01111 + 0.07917\,\sigma^{-2} - 0.1799\,\sigma^{-3} + 0.1055\,\sigma^{-4} + 0.02461\,\sigma^{-5}$
$C_5' = 0.001058 - 0.004960\,\sigma^{-2} + 0.001653\,\sigma^{-3} - 0.07348\,\sigma^{-4} - 0.03281\,\sigma^{-5}$
$C_6' = 0.0001587 - 0.0007275\,\sigma^{-2} + 0.0003638\,\sigma^{-3} - 0.08630\,\sigma^{-4} + 0.4000\,\sigma^{-5}$
$C_7' = 0.00003848 - 0.0001714\,\sigma^{-2} + 0.0001142\,\sigma^{-3} + 0.006183\,\sigma^{-4} - 0.002897\,\sigma^{-5}$

References

1. Saito N, Ikeda Y (1951) The light scattering by non-spherical particles in solutions. J Phys Soc Jpn 6:305–308. https://doi.org/10.1143/Jpsj.6.305
2. Terao K, Terao Y, Teramoto A, Nakamura N, Fujiki M, Sato T (2001) Temperature and solvent dependence of stiffness of poly{n-hexyl-[(s)-3-methylpentyl]silylene} in dilute solutions. Macromolecules 34:4519–4525. https://doi.org/10.1021/ma010212w

3. Terao K, Murashima M, Sano Y, Arakawa S, Kitamura S, Norisuye T (2010) Conformational, dimensional, and hydrodynamic properties of amylose tris(*n*-butylcarbamate) in tetrahydrofuran, methanol, and their mixtures. Macromolecules 43:1061–1068. https://doi.org/10.1021/ma902200z
4. Yamakawa H, Yoshizaki T (2016) Helical wormlike chains in polymer solutions, 2nd ed. Springer, Berlin. https://doi.org/10.1007/978-3-662-48716-7
5. Nakamura Y, Norisuye T (2008) Brush-like polymers. In: Borsali R, Pecora R (eds) Ch. 5. Springer Netherlands, p 235–286. https://doi.org/10.1007/978-1-4020-4465-6
6. Glatter O, Kratky O (1982) Small angle x-ray scattering. Academic Press, London.
7. Arakawa S, Terao K, Kitamura S, Sato T (2012) Conformational change of an amylose derivative in chiral solvents: Amylose tris(n-butylcarbamate) in ethyl lactates. Polym Chem 3:472–478. https://doi.org/10.1039/c1py00432h
8. Livsey I (1987) Neutron-scattering from concentric cylinders - intraparticle interference function and radius of gyration. J Chem Soc Faraday Trans 2 83:1445–1452. https://doi.org/10.1039/F29878301445
9. Nakamura Y, Norisuye T (2004) Scattering function for wormlike chains with finite thickness. J Polym Sci Part B Polym Phys 42:1398–1407. https://doi.org/10.1002/polb.20026
10. Nagai K (1972) Theory of light scattering by an isotropic system composed of anisotropic units with application to the porod–kratky chain. Polym J 3:67–83. https://doi.org/10.1295/polymj.3.67
11. Sakurai K, Ochi K, Norisuye T, Fujita H (1984) Stiff-chain behavior of poly(terephthalamide-p-benzohydrazide) in dimethylsulfoxide. Polym J 16:559–567. https://doi.org/10.1295/polymj.16.559
12. Terao K, Mays JW (2004) On-line measurement of molecular weight and radius of gyration of polystyrene in a good solvent and in a theta solvent measured with a two-angle light scattering detector. Europ Polym J 40:1623–1627. https://doi.org/10.1016/j.eurpolymj.2004.03.010
13. Nakamura Y, Wan Y, Mays JW Iatrou H, Hadjichristidis N (2000) Radius of gyration of polystyrene combs and centipedes in solution. Macromolecules 33:8323–8328. https://doi.org/10.1021/ma0007076
14. Park SW, Chang TH, Park IH (1991) Static and hydrodynamic size of polystyrene coils in various solvents. Macromolecules 24:5729–5731. https://doi.org/10.1021/ma00020a038
15. Schulz VGV, Baumann H (1968) Thermodynamisches verhalten, expansionskoeffizient und viskositätszahl von polystyrol in tetrahydrofuran. Makromol Chem 114:122–138. https://doi.org/10.1002/macp.1968.021140109
16. Bruce BJ, Pecora R (1976) Dynamic Light Scattering. Dober; New York.
17. Sugiyama M, Nakamura Y, Norisuye T (2008) Dilute-solution properties of polystyrene polymacromonomer having side chains of over 100 monomeric units. Polym J 40:109–115. https://doi.org/10.1295/polymj.PJ2007157
18. Fujita H (1975) Foundation of ultracentrigugal analysis. John Wiley & Sons, New York.
19. Murakami H, Norisuye T, Fujita H (1980) Dimensional and hydrodynamic properties of poly(hexyl isocyanate) in hexane. Macromolecules 13:345–352. https://doi.org/10.1021/ma60074a026
20. Arai T, Abe F., Yoshizaki T, Einaga Y, Yamakawa H (1995) Excluded-volume effects on the hydrodynamic radius of oligo- and polystyrenes in dilute solution. Macromolecules 28:3609–3616. https://doi.org/10.1021/ma00114a014
21. Yamada T, Yoshizaki T, Yamakawa H (1992) Transport coefficients of helical wormlike chains. 5. Translational diffusion coefficient of the touched-bead model and its application to oligo- and polystyrnenes. Macromolecules 25:377–383. https://doi.org/10.1021/ma00027a058
22. Kirkwood JG, Riseman J (1948) The intrinsic viscosities and diffusion constants of flexible macromolecules in solution. J Chem Phys 16:565–573. https://doi.org/10.1063/1.1746947
23. Flory PJ (1953) Principles of polymer chemistry. Cornel University Press, Ithaka.
24. Abe F, Einaga Y, Yoshizaki T, Yamakawa H (1993) Excluded-volume effects on the mean-square radius of gyration of oligo- and polystyrenes in dilute solutions Macromolecules 26:1884–1890. https://doi.org/10.1021/ma00060a014

25. Abe F, Einaga Y, Yamakawa H (1993) Excluded-volume effects on the intrinsic viscosity of oligomers and polymers of styrene and isobtylene. Macromolecules 26:1891–1897. https://doi.org/10.1021/ma00060a015
26. Miyaki M, Einaga Y, Fujita H (1978) Excluded-volume effects in dilute polymer solutions. 7. Very high molecular weight polystyrene in benzene and cyclohexane. Macromolecules 11:1180–1186. https://doi.org/10.1021/ma60066a022
27. Einaga Y, Miyaki M, Fujita H (1979) Intrinsic viscosity of polystyrene. J Polym Sci: Polym Phys Ed 17:2103–2109. https://doi.org/10.1002/pol.1979.180171205
28. Einaga Y, Koyama H, Konishi T, Yamakawa H (1989) Intrinsic viscosity of oligo- and polystyrene. Macromolecules 22:3419–24. https://doi.org/10.1021/ma00198a037
29. Konishi T, Yoshizaki , Saito T, Einaga Y, Yamakawa H (1990) Mean-square radius of gyration of oligo- and polystyrene in dilute solutions. Macromolecules 23:290–297. https://doi.org/10.1021/ma00203a050
30. Kratky O, Porod G (1949) Rontgenuntersuchung geloster fadenmolekule Recl Trav Chim Pays Bas 68:1106–1122. https://doi.org/10.1002/recl.19490681203
31. Benoit H, Doty P (1953) Light scattering from non-Gaussian chains J Phys Chem 57:958–963. https://doi.org/10.1021/j150510a025
32. Domb C, Barrett AJ (1976) Universality approach to the expansion factor of a polymer chain. Polymer 17:179–184. https://doi.org/10.1016/0032-3861(76)90096-3
33. Shimada J, Yamakawa H (1986) Statistical mechanics of helical worm-like chains. XV. Excluded-volume effects. J Chem Phys 85:591–600. https://doi.org/10.1063/1.451853
34. Terao K, Takeo Y, Tazaki M, Nakamura Y, Norisuye T (1999) Polymacromonomer consisting of polystyrene. Light scattering characterization in cyclohexane. Polym J 31:193–198. https://doi.org/10.1295/polymj.31.193
35. Terao K, Nakamura Y, Norisuye T (1999) Polymacromonomer consisting of polystyrene 2. Chain dimensions and stiffness in cyclohexane and toluene. Macromolecules 32:711–716. https://doi.org/10.1021/ma9816517
36. Hokajo T, Terao K, Nakamura Y, Norisuye T (1999) Solution properties of polymacromonomers consisting of polystyrene V. Effect of side chain length on chain stiffness Polym J 33:481–485. https://doi.org/10.1295/polymj.33.481
37. Yamakawa H, Fujii M (1974) Intrinsic viscosity of wormlike chains. Determination of the shift factor. Macromolecules 7:128–135. https://doi.org/10.1021/ma60037a024
38. Yamakawa H, Yoshizaki T (1980) Transport coefficients of helical wormlike chains. 3. Intrinsic viscosity. Macromolecules 13:633–643. https://doi.org/10.1021/ma60075a029
39. Roitman DB, Wessling RA, McAlister M (1993) Characterization of poly(p-phenylene-cis-benzobisoxazole) in methanesulfonic acid. Macromolecules 26:5174–5184. https://doi.org/10.1021/ma00071a030
40. Yoshizaki T, Nitta I, Yamakawa H (1988) Transport coefficients of helical wormlike chains. 4. Intrinsic viscosity of the touched-bead model. Macromolecules 21:165–171. https://doi.org/10.1021/ma00179a033
41. Barrett AJ (1984) Intrinsic viscosity and friction coefficients for an excluded volume polymer in the Kirkwood approximations. Macromolecules 17:1566–1572. https://doi.org/10.1021/ma00138a024
42. Terao K, Hokajo T, Nakamura Y, Norisuye T (1999) Solution properties of polymacromonomers consisting of polystyrene. 3. Viscosity behavior in cyclohexane and toluene. Macromolecules 32:3639–3694. https://doi.org/10.1021/ma990091o
43. Yamakawa H, Fujii M (1973) Translational friction coefficient of wormlike chains. Macromolecules 6:407–415. https://doi.org/10.1021/ma60033a018
44. Norisuye T, Motowoka M, Fujita H (1979) Wormlike chains near the rod limit: translational friction coefficient. Macromolecules 12:320–323. https://doi.org/10.1021/ma60068a032
45. Terao K, Hayashi S, Nakamura Y, Norisuye T (2000) Solution properties of polymacromonomers consisting of polystyrene 4. Translational diffusion coefficient in cyclohexane at the Θ temperature. Polym. Bull. 44:309–316. https://doi.org/10.1007/s002890050607

46. Hokajo T, Hanaoka Y, Nakamura Y, Norisuye T (2005) Translational diffusion coefficient of polystyrene polymacromonomers. Dependence on side-chain length. Polym J 37:529–534. https://doi.org/10.1295/polymj.37.529
47. Zimm BH, Stockmayer WH (1949) The dimensions of chain molecules containing branches and rings. J Chem Phys 17:1301–1314. https://doi.org/10.1063/1.1747157
48. Berry GC, Orofino TA (1964) Branched polymers. III. Dimensions of chains with small excluded volume. J Chem Phys 40:1614–1621. https://doi.org/10.1063/1.1725369
49. Cotts PM, Guan Z, McCord E, McLain S (2000) Novel branching topology in polyethylenes As revealed by light scattering and ^{13}C NMR. Macromolecules 33:6945–6952. https://doi.org/10.1021/ma000926r
50. Hadjichristidis N, Xenidou M, Iatrou H, Pitsikalis M, Poulos Y, Avgeropoulos A, Sioula S, Paraskeva S, Velis G, Lohse DJ, Schulz DN, Fetters LJ, Wright PJ, Mendelson RA, Garcia-Franco CA, Sun T, Ruff CJ (2000) Well-defined, model long chain branched polyethylene. 1. Synthesis and characterization. Macromolecules 33:2424–2436. https://doi.org/10.1021/ma991670w
51. Zimm BH, Kilb RW (1959) Dynamics of branched polymer molecules in dilute solution. J Polym Sci 37:19-42. https://doi.org/10.1002/pol.1959.1203713102
52. Noda I, Horikawa T, Kato T, Fujimoto T, and Nagasawa M (1970) Solution properties of comb-shaped polystyrenes. Macromolecules 3:795–799. https://doi.org/10.1021/ma60018a015
53. Strazielle C, Herz J (1977) Synthese synthese et etude physicochimique de polystyrenes ramifies en "etoile" et en "peigne". Europ Polym J 13:223–233. https://doi.org/10.1016/0014-3057(77)90118-5
54. Roovers J (1999) Dilute solution properties of regular stara polymers. In Star and Hyperbranched Polymers Mishra M, Kobayashi S Eds Marcel Dekker, New York.
55. Provencher SW (1982) CONTIN: A general purpose constrained regularization program for inverting noisy linear algebraic and integral equations. Comp Phys Commun 27:229–742. https://doi.org/10.1016/0010-4655(82)90174-6
56. Kanao M, Matsuda Y, Sato T (2003) Characterization of Polymer Solutions Containing a Small Amount of Aggregates by Static and Dynamic Light Scattering. Macromolecules 36:2093–2102. https://doi.org/10.1021/ma0213899

Chapter 4
Microstructures

4.1 Overview

The importance of precise control of polymer properties continues to increase in response to social demands for higher performance and functionality of polymeric materials. The primary structures of the polymers have significant effects on the polymer properties. Quantitative and reliable information on these primary structures can be obtained by various instrumental analyses such as nuclear magnetic resonance (NMR) spectroscopy. In this chapter, microstructural characterizations of polymers were described based on NMR spectra. Stereochemistry is one of the important factors to determine polymer properties. At first, the principle for the observation of stereochemistry in vinyl polymers was introduced. The importance of the measurement conditions such as solvent and temperature was described. Then, the method to assign the signals in terms of stereosequences was explained with polymerization statistics based on Bernoullian and Markovian statistics. Monomer sequence is another important factor to determine polymer properties. The definitions concerning monomer sequences were first described. The method to determine monomer reactivity ratios was explained. It is difficult to conduct a quantitative analysis of monomer sequences because NMR signals are split by both monomer sequences and stereosequences. Therefore, a chemometric approach to obtain the information on monomer sequences without assigning the individual signals was introduced. Chain-end groups are important from the viewpoints not only of polymer properties but also of polymerization mechanisms. Chain-end analyses of polymers prepared by radical and anionic polymerizations were described to figure out their polymerization mechanisms, respectively.

Authors: T. Hirano (hirano@tokushima-u.ac.jp) (Sect. 4.2) and A. Takano (atakano@chembio.nagoya-u.ac.jp) (Sect. 4.3).

Y. Nakamura et al., *Characterization of Polymers in Solution*,
https://doi.org/10.1007/978-981-96-6677-5_4

4.2 Nuclear Magnetic Resonance

NMR spectroscopy is a powerful tool for the microstructural characterization of polymers. Solution-state NMR spectroscopy is indispensable for analyzing polymer microstructures, including stereochemistry, monomer sequence, and chain-end structures.

4.2.1 *Stereochemical Analysis by NMR*

Stereochemical analysis of vinyl polymers by ^{1}H NMR was first reported for poly(methyl methacrylate) (PMMA) by three research groups at the beginning of the 1960s, independently [1–3]. The assignments were made, considering the magnetic equivalency of the in-chain methylene protons. Figure 4.1 shows the smallest unit for stereochemistry in vinyl polymers depicted in the rotated Fischer projection. The methylene protons (H_a and H_b) in the *m* diad are nonequivalent. The H_a is located between two methyl groups, whereas the H_b is located between two methoxycarbonyl groups. Accordingly, the signals of these protons are observed at different chemical shifts with splitting due to geminal spin-coupling, resulting in AB quartet spectral pattern. However, the methylene protons (H_c and H_d) in the *r* diad are magnetically equivalent because the chemical structure superposes if rotated 180° around the methylene carbon. Accordingly, the signals of these protons are observed at the same chemical shift. The mass fractions of the *m* and *r* diads, so-called diad tacticity, can be determined using integral intensities. Such a ^{1}H NMR spectroscopic characteristics is the absolute measure of the stereochemical configuration of the vinyl polymers. Note that the use of the terms "*meso*" and "*racemo*" is no longer recommended because of the potential for confusion with the similar terms, "*meso*" and "*racemi*", used in organic chemistry to describe small molecules [4].

Figure 4.2 shows continuous wave (CW) NMR spectra of PMMAs prepared with benzoyl peroxide (BPO) or Grignard reagents. The ^{1}H signals of the α-methyl protons show splitting due to the triad stereochemistry as depicted in Fig. 4.3. Therefore, triad tacticity can be determined using the integral intensities. However, the assignment cannot be achieved only from the signal pattern and intensities. One of the methods for assigning the signals is to use the relationship between diad-triad stereosequences as follows [5]:

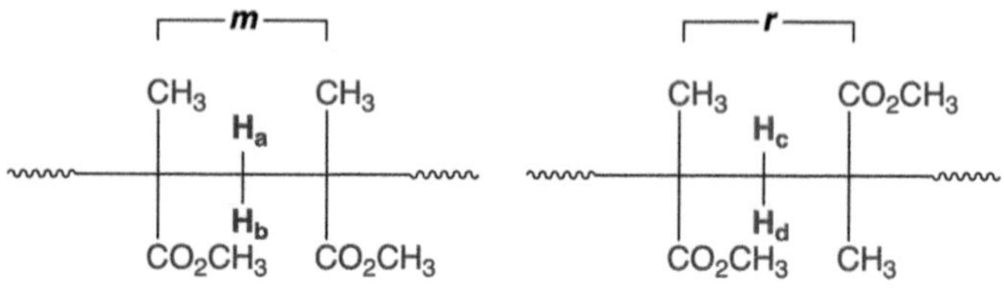

Fig. 4.1 Diad stereochemistry in PMMA

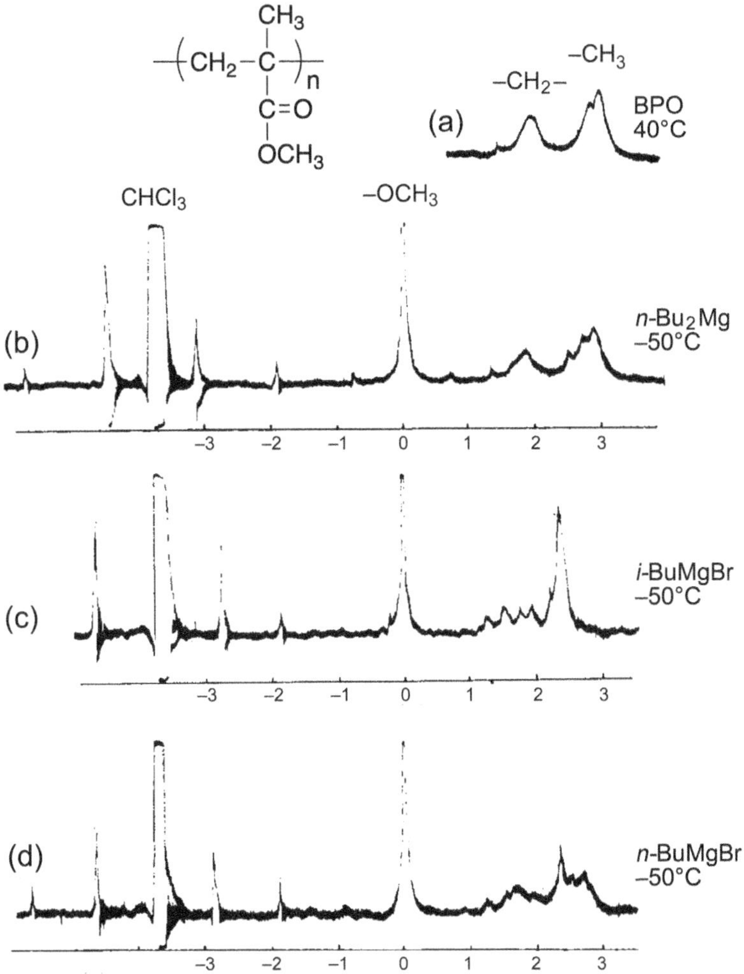

Fig. 4.2 56.4MHz ^{1}H NMR spectra of PMMAs prepared with **(a)** BPO, **(b)** $(n\text{-Bu})_2$Mg, **(c)** *i*-BuMgBr and **(d)** *n*-BuMgBr, respectively ($CHCl_3$, 20 °C) [2]

$$[m] = [mm] + \frac{1}{2}[mr]$$

$$[r] = [rr] + \frac{1}{2}[mr]$$

The ^{1}H NMR signals of the in-chain methylene groups of PMMA prepared with *i*-BuMgBr at –50 °C show a typical AB quartet pattern, indicating that the polymer has *m* diad abundantly (*cf*. Fig. 4.2). The signal of the α-methyl groups at the lowest magnetic field is observed strongly. Consequently, the signal can be assigned to *mm* triad.

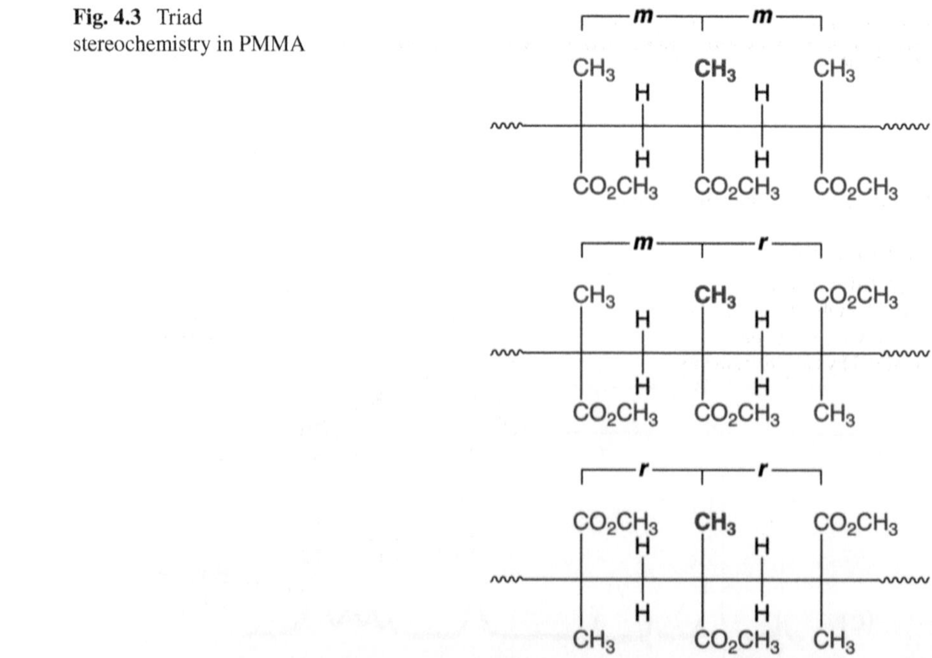

Fig. 4.3 Triad stereochemistry in PMMA

The increase in the resonance frequency enhances peak separation and resolution, resulting in splitting of the signals due to longer configurational sequences. For example, the 500 MHz ^{1}H signals of methylene protons exhibit splitting due to tetrad stereosequences (Fig. 4.4). Similarly, the signals of the α-methyl groups exhibit splitting due to pentad stereochemistry. Note that the signals of the in-chain methylene groups reflect the information on the stereochemistry of even-numbered monomeric units, whereas those of the α-methyl groups reflect the information on the stereochemistry of odd-numbered monomeric units.

The line width of NMR signal depends on the molecular motion: increased molecular motion narrows the line width. Accordingly, NMR signals of polymers are generally broadened compared with those of small molecules. Increased temperature enhances the molecular motion of polymers, resulting in improvement of signal separation and resolution (Fig. 4.5). Consequently, the NMR measurements of polymers are essentially desirable to be conducted at elevated temperatures as high as possible.

The measurement solvent is another important factor affecting spectral patterns. For example, the signal of the methoxy groups of PMMA measured in $CDCl_3$ shows no splitting due to stereochemistry. However, the signal of the methoxy groups measured in benzene-d_6 exhibit splitting due to triad stereochemistry, probably because of the magnetic anisotropy effect of the aromatic solvent (Fig. 4.6). It is noteworthy that methoxy groups are three bond lengths and therefore the furthest distance away from the stereogenic centers. A similar tendency is observed for

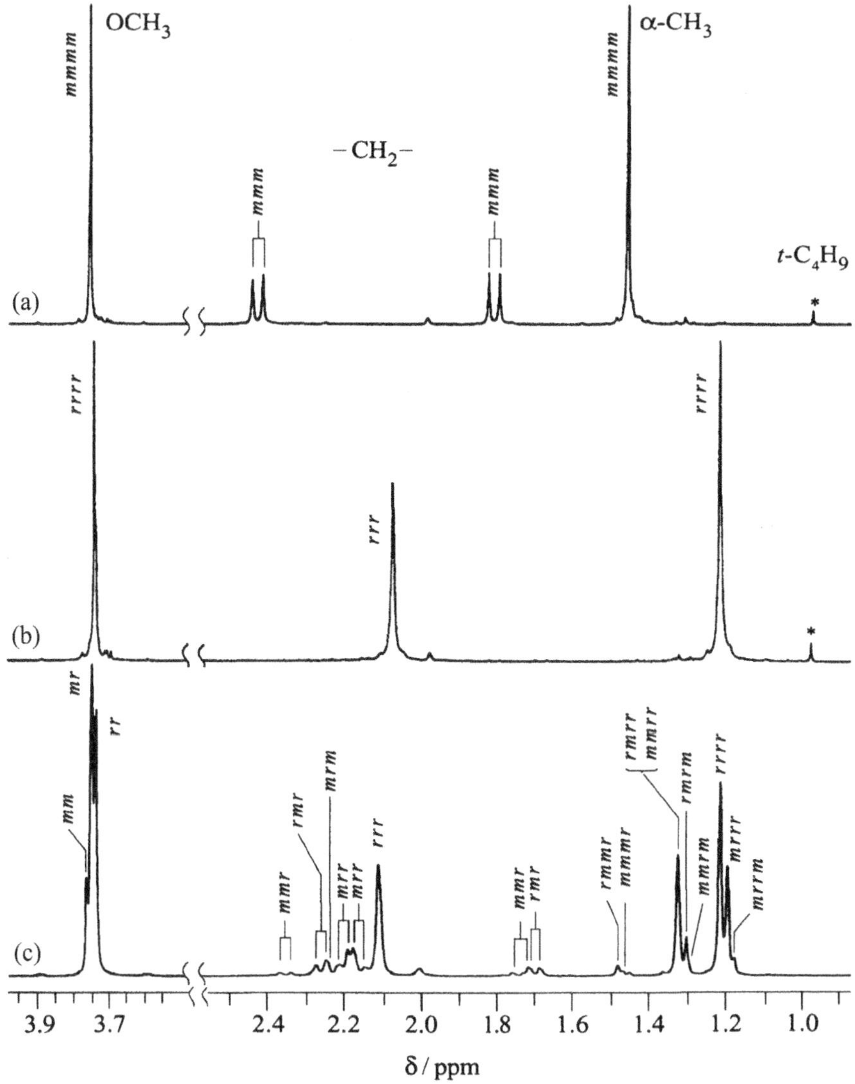

Fig. 4.4 500 MHz ^{1}H NMR spectra of (**a**) isotactic, (**b**) syndiotactic and (**c**) atactic PMMAs measured in nitrobenzene-d_5 at 110 °C. Polymer A was prepared with *tert*-C_4H_9MgBr in toluene at –78 °C, polymer B was derived from poly(trimethylsilyl methacrylate) prepared with *tert*-C_4H_9Li/bis(2,6-di-*tert*-butylphenoxy)methylaluminum in toluene at –95 °C, and polymer C was prepared with AIBN at 60 °C [6]

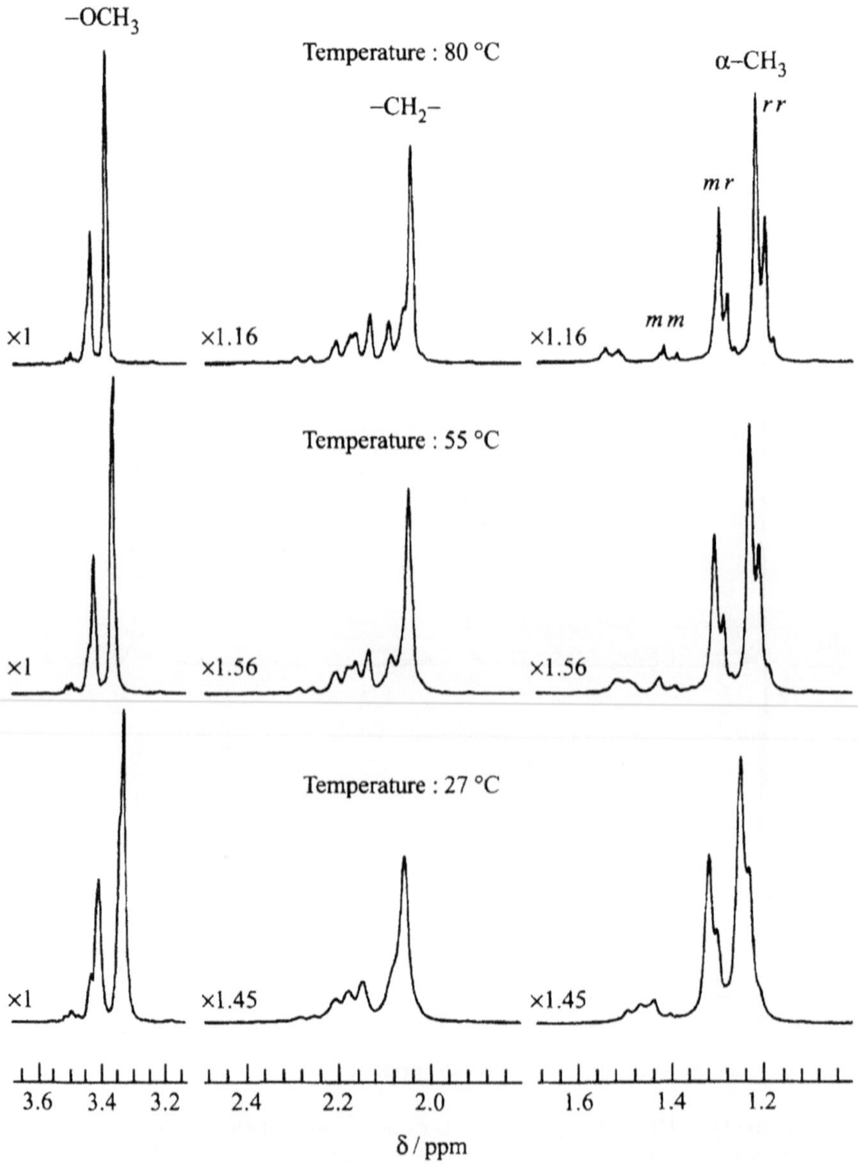

Fig. 4.5 Temperature dependence of ^{1}H NMR signals of radically prepared PMMA measured in benzene-d_6 [6]

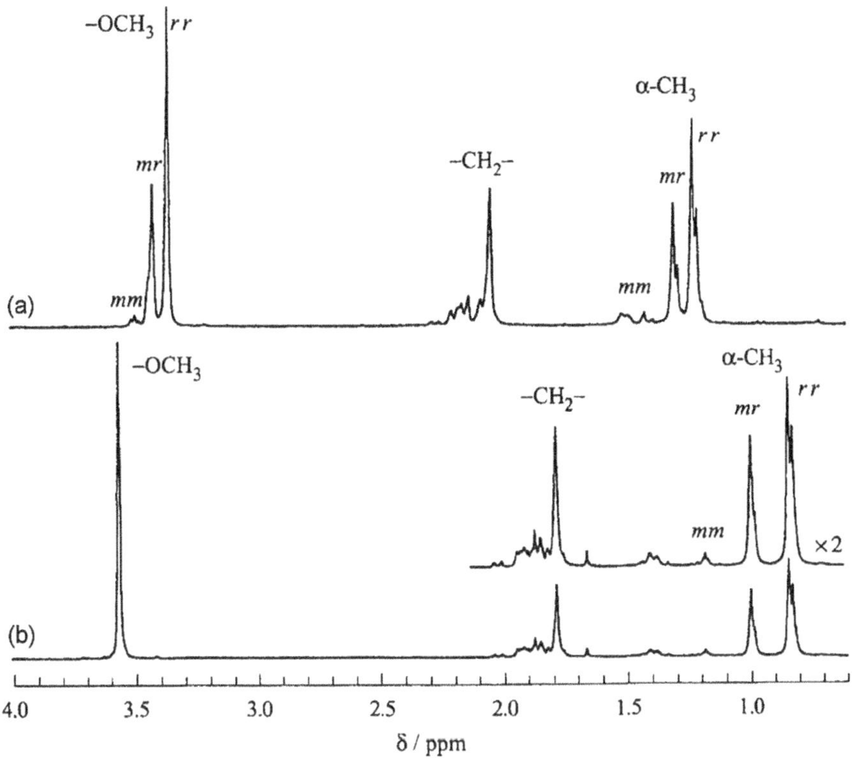

Fig. 4.6 500 MHz ^{1}H NMR spectra of radically prepared PMMA measured (**a**) in benzene-d_6 at 80 °C and (**b**) in $CDCl_3$ at 55 °C, respectively [6]

poly(*N*-vinyl-2-pyrrolidone) (PVP), where the signals of the methylene groups adjacent to carbonyl groups exhibited splitting due to triad stereochemistry, even though the methylene groups are three bond lengths from the stereogenic centers (Fig. 4.7) [7]. However, decreased temperature improves the signal separation and resolution in this case, probably because the effect of hydrogen bonding by D_2O surpasses the peak-broadening effect of decreased temperature.

Besides ^{1}H nuclei, other NMR active nuclei can be used for analysis of stereochemistry of polymer. ^{13}C nuclei is most probable nuclei, because overlapping of signals is generally reduced due to a wider range of spectral width (^{13}C: 200 ppm, ^{1}H: 10 ppm). Furthermore ^{1}H nuclei are usually broad-band decoupled to improve signal-noise ratio, because of low sensitivity of ^{13}C nuclei, resulting in much simpler ^{13}C signals than ^{1}H signals. Consequently, longer stereosequnces, such as tetrad and pentad, can be analyzed using ^{13}C NMR spectra. The measurement solvent and temperature are important for the signal separation and resolution also in ^{13}C NMR spectra (Fig. 4.8). For example, the signals of the carbonyl groups of PMMA show splitting due to pentad stereosequences, partially heptad stereosequences. However,

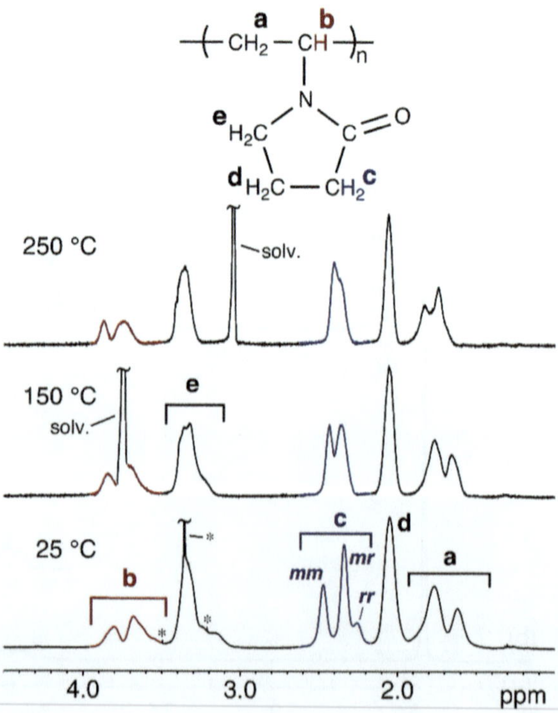

Fig. 4.7 500 MHz ^{1}H NMR spectra of PVP measured in D_2O at temperatures in the range of 25–250 °C. The asterisk (*) denotes impurities

the splitting due to the pentad tacticity as measured in toluene-d_8 at 105 °C is much more remarkable than that in $CDCl_3$ at 55 °C.

^{19}F NMR is also useful for the stereochemical analysis of fluorine-containing polymers, in particular polymers derived from α-fluoroacrylic acid derivatives [8–10]. The signals show splitting due to at least triad stereosequences (Fig. 4.9). The importance of the measurement solvent and the temperature remains even for ^{19}F NMR spectra.

4.2.2 *Stereosequence Analysis with Polymerization Statistics*

Even though NMR signals exhibit splitting due to stereosequences, the assignment of NMR signals is difficult. The assignment is usually conducted by comparing the observed intensities of signals with the intensities calculated with the polymerization statistics based on Bernoullian and first-order Markovian statistics as mentioned below.

Bernoullian statistics define a process in which the probability that a phenomenon takes place does not depend on the phenomena which happened before. In the case of stereochemistry in polymerization, the probability of the formation of *m* diad (P_m) does not depend on the chain-end stereochemistry. The probability of the formation

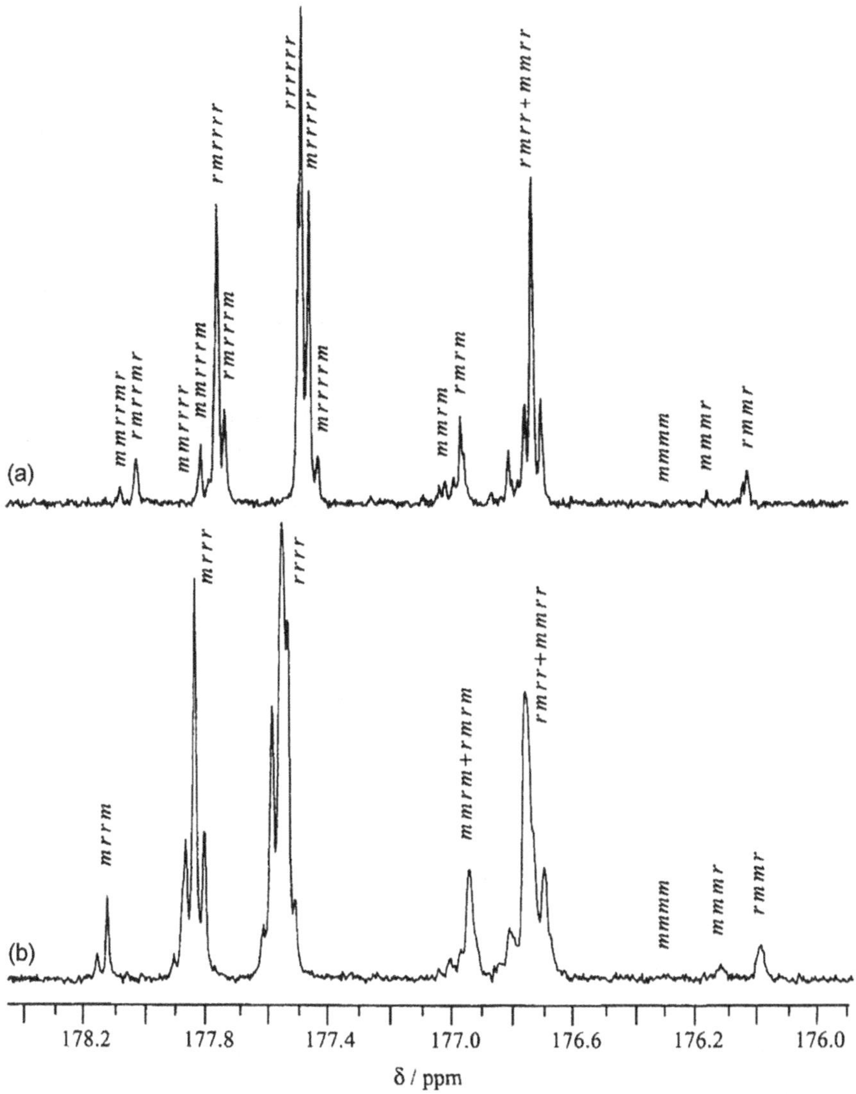

Fig. 4.8 125MHz ^{13}C NMR spectra of the carbonyl carbons of radically prepared PMMA measured (**a**) in toluene-d_8 at 105 °C and (**b**) in $CDCl_3$ at 55 °C, respectively [6]

of r diad (P_r) is $1 - P_m$, because there are only two stereochemistries in diad. If Bernoullian statistics are to hold in the stereochemistry of vinyl polymerization, the fraction of triad stereosequences can be described using one parameter (P_m) as follows:

$$[mm] = P_m^2,\ [mr] = 2P_mP_r = 2P_m(1 - P_m),\ [rr] = P_r^2 = (1 - P_m)^2$$

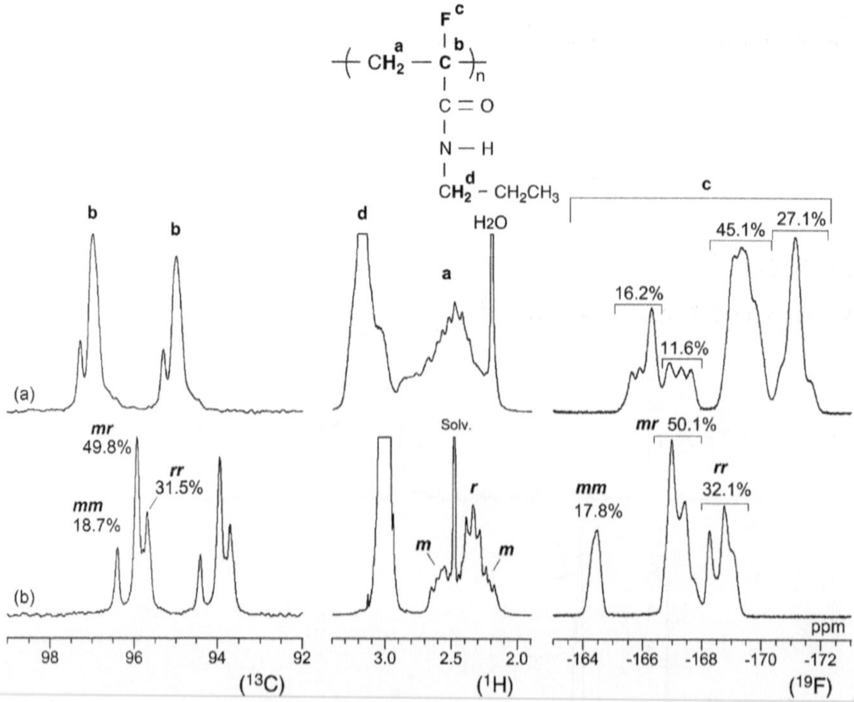

Fig. 4.9 ^{1}H, ^{13}C and ^{19}F NMR spectra of poly(*N*-*n*-propyl-α-fluoroacrylamide) measured (**a**) in $CDCl_3$ at 55 °C and (**b**) DMSO-d_6 at 100 °C, respectively [10]

The fraction of *mr* triad is given double statistical weighing because *mr* and *rm* triads cannot be distinguished. Longer stereosequences, such as tetrad, can be calculated similarly as follows:

$$[mmm] = P_m^3,\ [mmr] = 2P_m^2(1 - P_m),\ [rmr] = P_m(1 - P_m)^2,$$
$$[mrm] = P_m^2(1 - P_m),\ [mrr] = 2P_m(1 - P_m)^2,\ [rrr] = (1 - P_m)^3$$

Therefore, stereochemical fractions can be calculated if diad tacticity was determined.

When the probability of the formation of *m* diad depends on the chain-end diad, the fraction of stereosequences cannot be described by a single parameter. In that case, the first-order Markovian statistics must be introduced. The probabilities of the formation of *m* diad by *m*-ended and *r*-ended active species can be described as $P_{m/m}$ and $P_{r/m}$, respectively (Fig. 4.10). Because the following relations should hold, the number of independent parameters is 2:

$$P_{m/m} + P_{m/r} = 1, P_{r/m} + P_{r/r} = 1$$

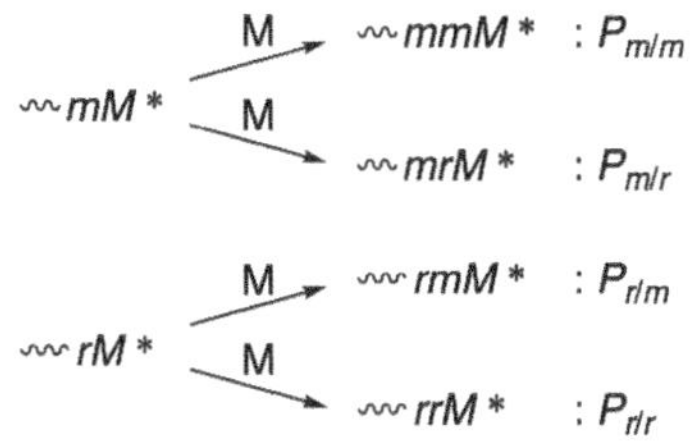

Fig. 4.10 Propagating reactions by *m*-ended and *r*-ended active chains and the probabilities of the formation of *m* and *r* diads through the propagating reactions. The asterisks (*) denote radical, anionic or cationic species

The parameters can be calculated from triad tacticity as follows:

$$P_{m/r} = \frac{[mr]}{2[mm] + [mr]},\ P_{r/m} = \frac{[mr]}{2[rr] + [mr]},$$

To verify if the tacticity distribution obeys the first-order Markovian statistics, the tacticity data of orders higher than triad are required. For example, the fractions of tetrad stereosequences can be calculated from triad data and the parameters derived thereof as follows:

$$[mmm] = [mm]P_{m/m},\ [mmr] = [mm]P_{m/r} + [rm]P_{m/m},\ [rmr] = [rm]P_{m/r},$$
$$[mrm] = [mr]P_{r/m},\ [mrr] = [mr]P_{r/r} + [rr]P_{r/m},\ [rrr] = [rr]P_{r/r}$$

In Bernoullian statistics, the following relationship between triads are found:

$$[mr]^2 = [2P_m(1 - P_m)]^2 = 4P_m^2(1 - P_m)^2 = 4[mm]\,[rr]$$

Thus,

$$\frac{4[mm][rr]}{[mr]^2} = 1$$

Furthermore, if the probability of the formation of *m* diad is independent of the chain-end diad ($P_{m/m} = P_{r/m}$), the following equation holds.

$$P_{m/r} + P_{r/m} = 1$$

Therefore, if the signals are already assigned at the triad level, it is easy to check whether the polymerization process obeys Bernoullian statistics or not using the two relationships mentioned above.

Radical polymerization processes are said to obey Bernoullian statistics. Therefore, fractions of pentad stereosequences in the radically prepared PMMA were

Table 4.1 Comparison of observed and calculated fractions of pentad stereosequences for radically prepared PMMA [6]

Pentad	Observed[a]	Calculated[b]
mmmm	0.00	0.00
mmmr	0.01	0.01
rmmr	0.03	0.03
mmrm	0.01	0.01
mmrr	0.05	0.06
rmrm	0.06	0.06
rmrr	0.22	0.21
mrrm	0.03	0.03
mrrr	0.19	0.21
rrrr	0.40	0.38

[a] Determined from the carbonyl carbon signals in the spectrum measured in toluene-d_8

[b] The values were calculated from the probability P_{m} (= 0.214) assuming that the stereochemical sequence distribution in radically prepared PMMA obeys a Bernoullian statistics

calculated as summarized in Table 4.1. The Bernoullian parameter P_{m} was calculated to be 0.214 from the triad tacticity obtained from ^{1}H signals of the α-methyl groups of PMMA prepared with 2,2′-azobis(isobutyronitrile) (AIBN) in toluene at 60 °C (*mm* = 0.038, *mr* = 0.352, *rr* = 0.610). In this procedure, some pentads should have the same probability, such as *mmrm* & *mmmr*, *mmrr* & *rmrm*, and *rmrr* & *rrrm*. However, the signals may be apparently grouped in three, which are *mm*-, *mr*-, and *rr*-centered sequences as shown in Fig. 4.8, and thus the pairs of *mmrm* & *mmmr* and *rmrr* & *rrrm* are not interchangeable.

The pair of *mmrr* & *mrmr* is interchangeable. In that case, further comparison with calculated values for samples with different stereoregularities is required for the assignment, as described below. The observed values determined from the signal intensities of the carbonyl carbons of radically prepared PMMA measured in toluene-d_8 (Fig. 4.8b) are also summarized in Table 4.1. The observed values agreed well with the calculated values.

Figure 4.11 shows ^{13}C signals of heterotactic poly(allyl methacrylate) prepared by anionic polymerization [11]. In the carbonyl region, the *mr*-centered signals are most abundant. The signals due to *mrmr* pentad are observed at a lower magnetic field than those due to *mmrr* pentad. Therefore, the signals due to *mmrr* & *mrmr* can be assigned individually as shown in Fig. 4.8.

The methylene signals are composed of *m*-centered and *r*-centered signal groups with almost the same intensity, indicating that the content of *m* diad is close to 0.50. This is the characteristic feature in NMR spectra of heterotactic polymers. If the polymer with the *m* diad content of 0.50 obeys Bernoullian statistics, the polymer is a non-stereoregular polymer, *i.e.*, atactic polymer. The fraction of *mr* triad is calculated to be 0.50 with Bernoullian parameter P_m = 0.50. Accordingly, Bernoullian statistics cannot express the triad tacticity distribution in heterotactic

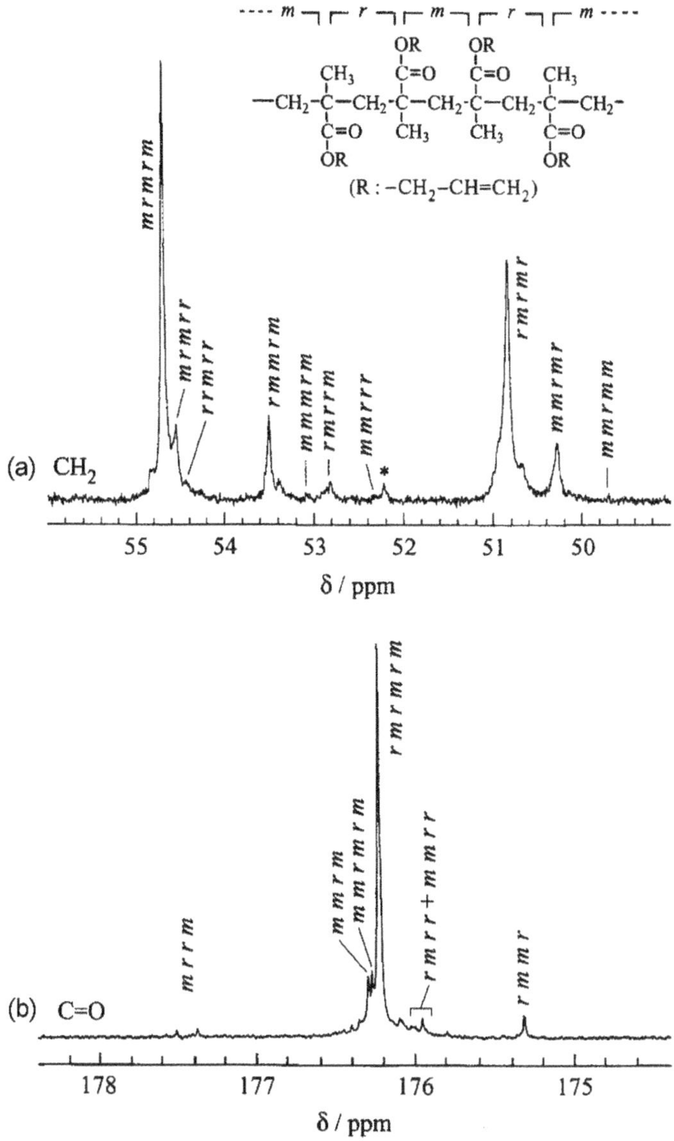

Fig. 4.11 125 MHz ^{13}C NMR signals of (**a**) the in-chain methylene and (**b**) carbonyl groups of heterotactic poly(allyl methacrylate) (*mr* triad content 95.8%) measured in $CDCl_3$ at 55 °C. The asterisk (*) denotes the signal of methylene carbons of terminal monomeric units [6]

Table 4.2 Some necessary relationships among stereosequence frequencies [5]

Diad	$[m] + [r] = 1$
Triad	$[mm] + [mr] + [rr] = 1$
Diad-triad	$[m] = [mm] + 1/2[mr]$ $[r] = [rr] + 1/2[mr]$
Triad-tetrad	$[mm] = [mmm] + 1/2[mmr]$ $[mr] = [mmr] + 2[rmr] = [mrr] + 2[mrm]$ $[rr] = [rrr] + 1/2[mrr]$
Tetrad-tetrad	Sum $= 1$ $[mmr] + 2[rmr] = [mrr] + 2[mrm]$
Tetrad-pentad	$[mmm] = [mmmm] + 1/2[mmmr]$ $[mmr] = [mmmr] + 2[rmmr] = [mmrm] + [mmrr]$ $[rmr] = 1/2[mrmr] + 1/2[rmrr]$ $[mrm] = 1/2[mrmr] + 1/2[mmrm]$ $[mrr] = 2[mrrm] + [mrrr] = [mmrr] + [rmrr]$ $[rrr] = [rrrr] + 1/2[mrrr]$
Pentad-pentad	$[mmmr] + 2[rmmr] = [mmrm] + [mmrr]$ $2[mrrm] + [mrrr] = [mmrr] + [rmrr]$

polymers, and heterotactic polymers require higher-order statistics, for example, first-order Markovian statistics, to analyze the stereosequence distribution. It is thus obvious that the analysis of longer stereosequences is important to know the higher level of stereoregularity.

To confirm the stereochemical assignments, it is advisable to check the necessary relationship among the stereosequences (Table 4.2), which is entirely independent of the statistics of the polymerization processes [5].

4.2.3 Monomer Sequence Analysis by NMR

Structural analysis of copolymer is very important from both scientific and industrial viewpoints because the copolymer properties depend not only on its composition but also on constitutional and configurational sequences. Solution-state NMR is particularly effective in characterization of vinyl copolymers. If a copolymer is produced only from nonprochiral monomers, such as comonomers of the type $CH_2{=}CX_2$, there exists a set of constitutional sequences shown in Table 4.3.

There are the necessary relationships among constitutional sequences as well as configurational sequences as follows:

$$
\begin{aligned}
[A] &= [AA] + 1/2[AB] \\
[B] &= [BB] + 1/2[AB] \\
[AA] &= [AAA] + 1/2[BAA] \\
[AB] &= 1/2[AAB] + [BAB] + 1/2[ABB] + [ABA] \\
[BB] &= [BBB] + 1/2[ABB]
\end{aligned}
$$

The average sequence lengths of the constitutional monomeric units can be determined from the diad distributions using the following relationships [12]:

$$
\bar{l}_A = \frac{[A]}{1/2[AB]} = \frac{[AA] + 1/2[AB]}{1/2[AB]}
$$
$$
\bar{l}_B = \frac{[B]}{1/2[AB]} = \frac{[BB] + 1/2[AB]}{1/2[AB]}
$$

where $\bar{l}_A$ and $\bar{l}_B$ denote the average sequence lengths of the monomeric units A and B, respectively.

The average sequence lengths depend on the chemical composition. Therefore, several "one-number" parameters, such as run number [13], persistence ratio [14, 15] and degree of randomness [16], have been used to describe the constitutional sequence distribution. For example, degree of randomness (R) is defined as follows:

$$
R = \frac{1}{\bar{l}_A} + \frac{1}{\bar{l}_B} = \frac{1/2[AB]}{[AA] + 1/2[AB]} + \frac{1/2[AB]}{[BB] + 1/2[AB]}
$$

The R values, ranging from 0.0 to 2.0, provide information on the distribution of monomer sequences (for a mixture of homopolymers $R = 0.0$, for random copolymers obeying Bernoullian statistics $R = 1.0$ and for alternating copolymers $R = 2.0$).

If a copolymer is prepared from prochiral monomers, such as comonomers of the type $CH_2{=}CHX$, or $CH_2{=}CXY$, configurational sequences are to be considered and there exist sets of stereosequences for diads and triads as shown in Table 4.4. The number of the sequences in n-ads increases largely with n.

Even though compositional analysis can be achieved by several methods, such as elemental analysis of copolymers and gas chromatography of polymerization mixtures, integral intensities of characteristic ^{1}H signals of the monomeric units easily provide quantitative chemical composition. For example, chemical

Table 4.3 Monomer sequences in copolymers poly(A-*co*-B) made of nonprochiral monomers A and B [5]

Diad	AA	AB or BA	BB
Triad	AAA BAA (or AAB) BAB		BBB ABB (or BBA) ABA

Table 4.4 Configurational sequences in copolymers poly(A-*co*-B) made of prochiral monomers A and B [5]

Diad		AA	AB (or BA)	BB
	m	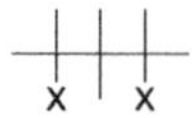	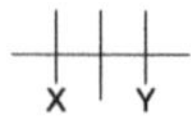	
	r	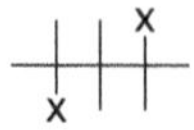	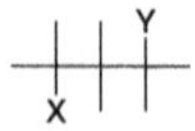	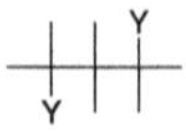
Triad[a]		AAA	BAA (or AAB)	BAB
	mm	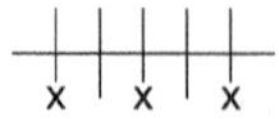	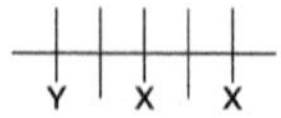	
	mr		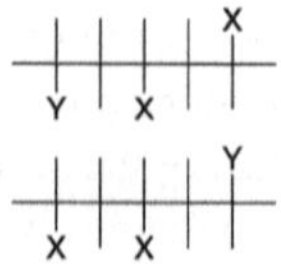	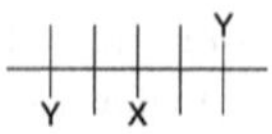
	rr	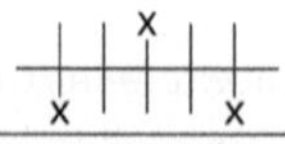	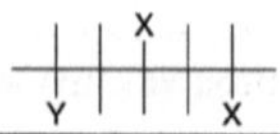	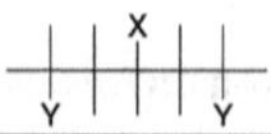

[a] Ten others with BBB, ABB (or BBA) and ABA triads

composition of copolymers comprising *N*-ethylacrylamide (NEAAm) and *N*-isopropylacrylamide (NIPAAm) can be determined from the intensities of the ^{1}H signals due to the *N*-C***H*₂** groups (**c**) in the NEAAm unit and *N*-C***H*** groups (**f**) in the NIPAAm unit (Fig. 4.12) [17]. A wide range of comonomer combinations, including MMA and ethyl acrylate (EA), is acceptable if characteristic signals for the comonomers are observed separately. It should be noted that the chemical compositions of the copolymer of MMA and EA are never determined by elemental analysis, because EA is a structural isomer of MMA.

The NMR signals of copolymers exhibit splitting not only due to monomer sequence but also due to stereosequence. Accordingly, the spectral pattern of copolymers is generally complicated. In particular, copolymers containing monomers of the same category, such as methacrylates and acrylates, show considerable overlap of the signals of the main chain backbone structures. Figure 4.13. shows ^{13}C NMR spectra of copolymers comprising MMA and *n*-butyl methacrylate (BMA) [6, 18]. In the spectra of stereoregular copolymers, splitting due to stereosequences is almost neglected. As a result, the signals split due to triad monomer sequences. However, a radically prepared copolymer showed a complicated spectral pattern. Even though the individual signals were assigned to triad monomer sequences with pentad stereosequences, the information on the monomer sequence cannot be obtained from the signal intensities. In such cases, multivariate analysis of NMR spectra is effective [19–22]. The details will be described in the next section.

The copolymerization reaction of two monomers (M_1 and M_2) is usually described by monomer reactivity ratios r_1 $(= k_{11}/k_{12})$ and r_2 $(= k_{22}/k_{21})$. Here, k_{ij} is the rate

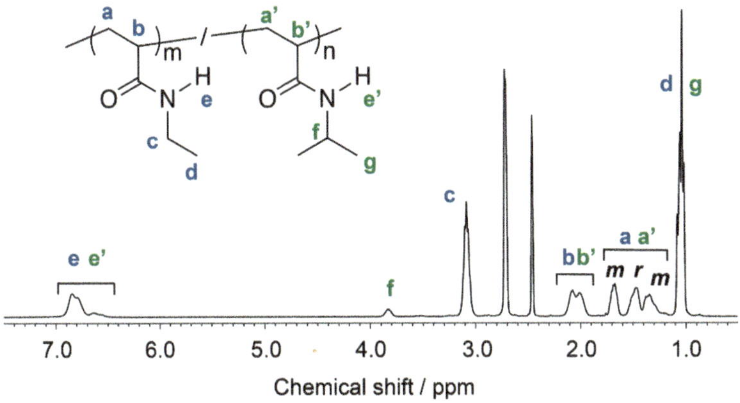

Fig. 4.12 400 MHz ^{1}H NMR spectrum of poly(NEAAm-*co*-NIPAAm) with NIPAAm content of 13.9 mol% measured in DMSO-d_6 at 150 °C. The copolymer was prepared with dimethyl 2,2′-azobisisobutyrate in *N*-ethylacetamide at –40 °C [17]

constant for the four kinds of propagating reactions (Fig. 4.14). For example, k_{12} is the rate constant for the reaction of M_1-ended active species with M_2 monomer.

The r_1 and r_2 can describe the sequence statistics of copolymerization. When the reactivity of the growing chain is independent of the nature of the monomeric unit at the growing chain-end ($r_1 = 1$, $r_2 = 1$), the monomer sequences obey Bernoullian statistics. When the reactivity depends on the growing chain-end monomeric unit, the monomer sequences obey first-order Markovian statistics. The monomer reactivity ratios are usually calculated by several methods, such as Finemann-Ross [23], Kelen-Tüdos [24], and nonlinear least-squares [25] methods. These methods require the preparation of multiple copolymers obtained at the early stage of copolymerization. Therefore, copolymerization reactions must be carried out by changing the monomer composition in feed.

When the mole fraction of each monomer sequence at diad level is obtained, the monomer reactivity ratios can be determined from a single sample of copolymers [26, 27]. For example, in ^{13}C NMR spectra of copolymers comprising MMA and acrylonitrile (AN), signals of C=O and CN groups in MMA and AN units, respectively, exhibited wider splitting due to triad monomer sequences than splitting due to stereosequences (Fig. 4.15). Therefore, the fractions of six triads of monomer sequences can be calculated by combining the chemical composition determined by ^{1}H NMR spectra.

The diad fractions can be derived from the triad fractions as shown in the following equations:

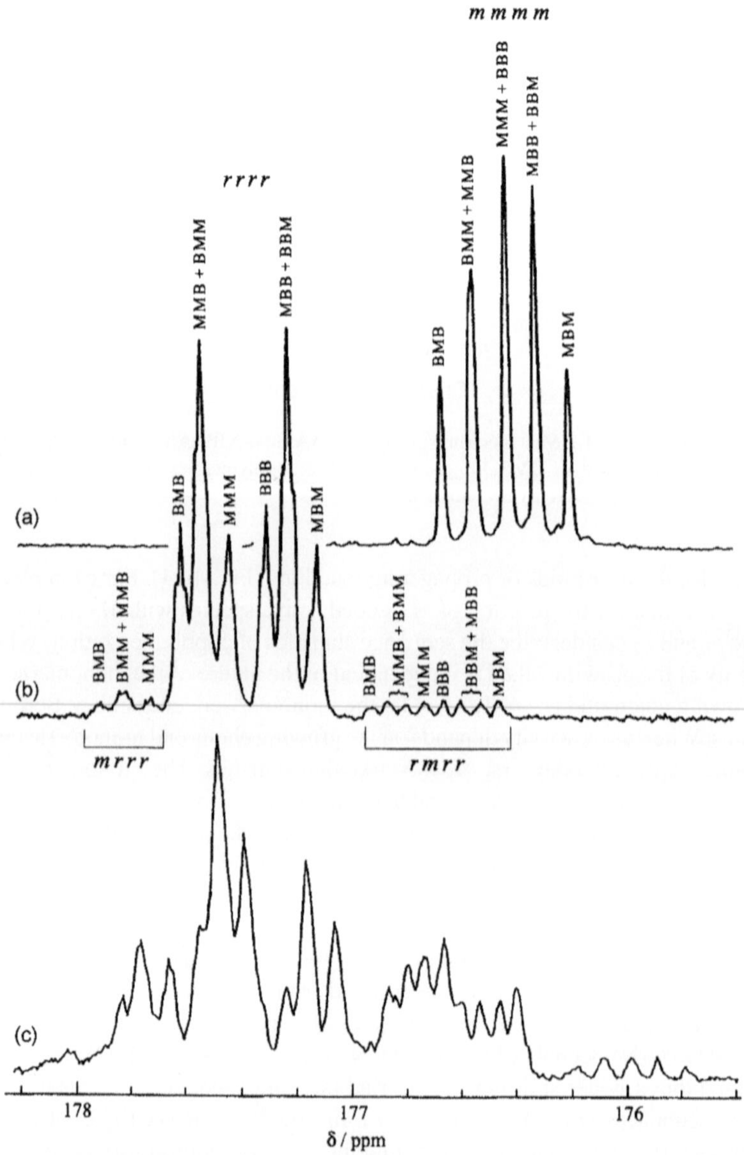

Fig. 4.13 ^{13}C NMR spectra of carbonyl groups of (**a**) isotactic (M:B = 49:51), (**b**) syndiotactic (M:B = 50:50) and (**c**) atactic (M:B = 59:41) poly(MMA-*co*-BMA)s, prepared with (**a**) *tert*-C_4H_9MgBr, (**b**) *tert*-$C_4H_9Li/(C_2H_5)_3Al$ and (**c**) AIBN, respectively, measured in chlorobenzene-d_5 at 115 °C. M and B represent MMA and BMA units in the copolymers, respectively [6]

Fig. 4.14 Four kinds of propagating reaction in copolymerization of M_1 and M_2 monomers. The asterisks (*) denote radical, anionic, or cationic species

$$\sim M_1^* + M_1 \xrightarrow{k_{11}} \sim M_1M_1^*$$

$$\sim M_1^* + M_2 \xrightarrow{k_{12}} \sim M_1M_2^*$$

$$\sim M_2^* + M_1 \xrightarrow{k_{21}} \sim M_2M_1^*$$

$$\sim M_2^* + M_2 \xrightarrow{k_{22}} \sim M_2M_2^*$$

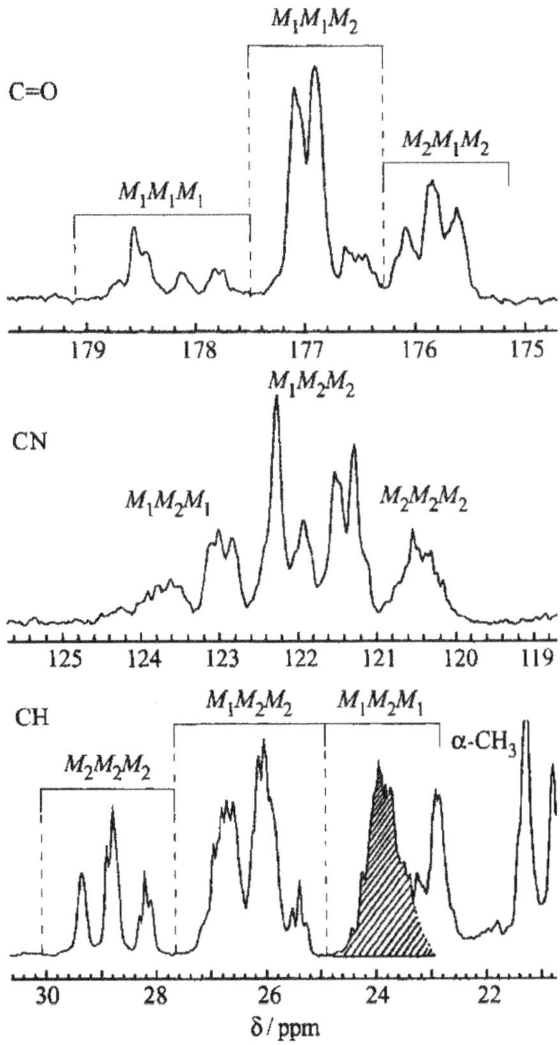

Fig. 4.15 100 MHz ^{13}C NMR spectra of poly(MMA-*co*-AN) with MMA content of 52.0 mol% measured in acetonitrile-d_3 at 70 °C [26]. In the CH region spectrum the hatched area is assigned to the $M_1M_2M_1$ triad according to the literature [28]

Table 4.5 Reactivity ratios r_i and r_{ij} for the copolymerization of MMA (M_1) and AN (M_2) in DMSO 50 °C

MMA composition		r_1	r_2	r_{11}	r_{21}	r_{12}	r_{22}
In feed	In copolymer						
0.206	0.375	1.55	0.34	1.44	1.56	0.34	0.34
0.347	0.520	1.41	0.34	1.24	1.55	0.34	0.33
0.548	0.671	1.27	0.38	1.13	1.52	0.38	0.39

$$[M_1M_1] = [M_1M_1M_1] + \frac{[M_1M_1M_2]}{2},$$
$$[M_1M_2] = [M_2M_1M_2] + [M_1M_2M_1] + \frac{[M_1M_1M_2]}{2} + \frac{[M_1M_2M_2]}{2},$$
$$[M_2M_2] = [M_2M_2M_2] + \frac{[M_1M_2M_2]}{2}$$

Consequently, the r_1 and r_2 values were calculated using the following equations. The calculated values for the copolymers prepared in different monomer compositions in feed ($[M_1]_0$ and $[M_2]_0$) are shown in Table 4.5. The monomer reactivity ratios depended on the monomer composition in feed. Particularly, r_1 decreased from 1.55 to 1.27 with an increase in MMA composition in feed from 0.206 to 0.548. This result indicated that the copolymerization of MMA and AN cannot be described with the terminal model based on first-order Markovian statistics [27–29].

$$r_1 = \frac{[M_2]_0}{[M_1]_0}\frac{2[M_1M_1]}{[M_1M_2]}$$
$$r_2 = \frac{[M_1]_0}{[M_2]_0}\frac{2[M_2M_2]}{[M_1M_2]}$$

When the reactivity of the growing chain depends on the natures of the ultimate and penultimate monomeric units, four parameters of reactivity ratios r_{11} ($= k_{111}/k_{112}$), r_{21} ($= k_{211}/k_{212}$), r_{12} ($= k_{122}/k_{121}$), and r_{22} ($= k_{222}/k_{221}$) are necessary to describe the chain statistics, where, for example, k_{111} and k_{211} are the rate constants for the following propagating reactions.

$$\sim M_1M_1^* + M_1 \xrightarrow{k_{111}} \sim M_1M_1M_1^*$$
$$\sim M_2M_1^* + M_1 \xrightarrow{k_{211}} \sim M_2M_1M_1^*$$

The penultimate-model reactivity ratios, r_{ij}, can be determined from triad monomer sequence distribution using the following equations:

$$r_{11} = \frac{[M_2]_0}{[M_1]_0}\frac{2[M_1M_1M_1]}{[M_1M_1M_2]}, \quad r_{21} = \frac{[M_2]_0}{[M_1]_0}\frac{[M_1M_1M_2]}{2[M_2M_1M_2]},$$
$$r_{12} = \frac{[M_1]_0}{[M_2]_0}\frac{[M_1M_2M_2]}{2[M_1M_2M_1]}, \quad r_{22} = \frac{[M_1]_0}{[M_2]_0}\frac{2[M_2M_2M_2]}{[M_1M_2M_2]}$$

The r_{ij} values are also summarized in Table 4.5. The r_{11} value is different from the r_{21} value for all samples. This is a clear indication of a significant penultimate effect characteristic of the MMA-ended radical. Furthermore, both values, particularly r_{11}, decreased with an increase in the MMA composition in feed. This suggests the possible existence of an antepenultimate effect in the reactivity of MMA-ended radicals. On the other hand, the values of r_{12} and r_{22} determined for each copolymer were close to each other, both increasing slightly with increasing MMA composition in feed. This suggests that there are little penultimate and antepenultimate effects for the AN-ended radicals in this copolymerization system.

4.2.4 Multivariate Analysis of NMR Spectra

The NMR signals of copolymers often exhibit overlapped splitting due to monomer sequences and stereosequences, and thus assignment of the individual peaks is troublesome as mentioned in the above section. Increasing the number of monomer components in copolymers makes the signals broader and more complicated.

Since chemometrics was established at the beginning of the 1970s by Kowalski [30, 31] and Wold [32], multivariate analyses, such as principal component analysis (PCA) and partial least-squares (PLS) regression, have been recognized as a powerful tool in the characterization of polymer structures.

For example, PCA and PLS regression of ^{13}C NMR spectra of 16 copolymers comprising MMA and *tert*-butyl methacrylate (TBMA), 2 corresponding homopolymers, and 9 blends of the homopolymers with different ratios were found to be useful for analyzing the monomer sequences in copolymers [20]. Figure 4.16 shows ^{13}C NMR expanded spectra of the carbonyl, backbone quaternary, and α-methyl carbons of PMMA, PTBMA, and poly(MMA-*co*-TBMA) with TBMA content of 55.5 mol%, which are the common structures in MMA and TBMA units, together with the corresponding PCA loading histogram. The loading histograms indicate the pseudo-signal reflecting each principal component in the bucket integral intensities for the three resonance regions of all sample polymers.

Positive loadings of PC1 were observed at the same chemical shift owing to the signals of PMMA, whereas negative ones were observed from PTBMA. The loading histograms indicated that PC1 reflected the chemical compositions of MMA and TBMA units in polymer samples. The PC2-positive loadings appeared at the signals of the two homopolymers. Conversely, the PC2-negative loadings were observed at the characteristic signals of copolymers. The prime characteristic of copolymers is the presence of the monomer sequences derived from cross-propagation reactions (*cf.* Fig. 4.14). Therefore, PC2 should reflect the heterogeneity of monomer sequences.

The fractions of diad monomer sequences (f_{MM}, f_{MT} and f_{TT}) can be calculated with the monomer reactivity ratios [33], if the copolymers were obtained in the early stage of polymerization. To predict fractions of diad monomer sequences, PLS regression was performed with a training set of copolymers obtained at low conversions. Figure 4.17 shows the relationship between theoretical and predicted

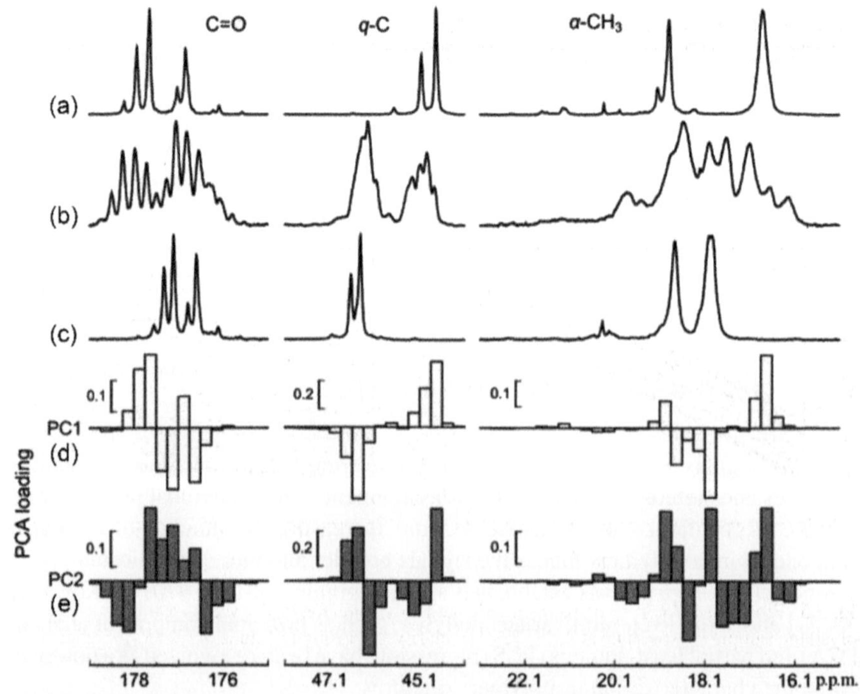

Fig. 4.16 ^{13}C NMR spectra of the carbonyl, backbone quaternary and α-methyl carbons of (**a**) PMMA, (**b**) poly(MMA-*co*-TBMA) and (**c**) poly(TBMA). The corresponding loading histograms for the (**d**) first (PC1) and (**e**) second (PC2) principal components of the ^{13}C NMR spectra [20]

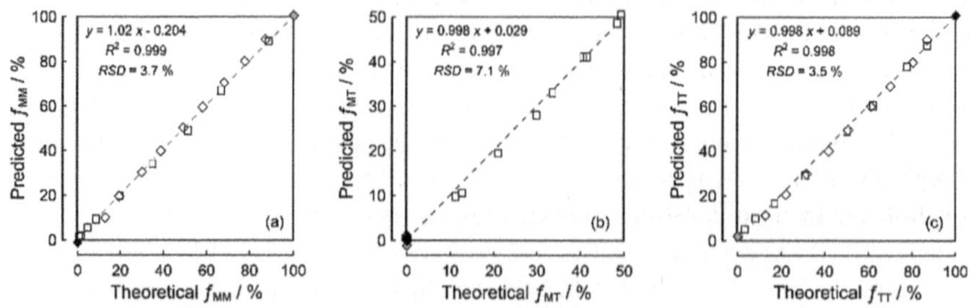

Fig. 4.17 Relationships of the fractions of (**a**) MM, (**b**) MT, and (**c**) TT diad monomer sequences, respectively, predicted by PLS regression with the theoretical values (◆: PMMA, ◆: poly(TBMA), ◇: blends, □: copolymers) [20]

fraction values. The predicted values were obtained by leave-one-out cross-validation method. The good fit between the theoretical and predicted values indicates that the fractions of the diad monomer sequences can be predicted with good accuracy and precision, even from spectra of unknown samples. Notably, the assignment of individual signals was not necessary in this method to predict the monomer sequence fractions.

4.2.5 Chain-End Analysis by NMR

In addition to the structural analysis of the in-chain structures such as constitutional and configurational sequences, analysis of chain-end groups of polymer chains is important because the chain-end structures not only reflect the polymerization mechanism but also influence the thermal properties. Figure 4.18 shows 750 MHz ^{1}H NMR spectrum of PMMA prepared with BPO in toluene. The expanded spectra inserted in the figure show signals assignable to chain-end groups and chain-end monomeric units. In the low magnetic field, signals of benzoyloxy group (Structure I, Fig. 4.19) derived from the initiator BPO (Fig. 4.18, range A) and aromatic group (Fig. 4.18, range B) were observed. The latter aromatic signals were further assigned into phenyl (Structure II) and benzyl (Structure III) groups. The phenyl group was formed by the addition of phenyl radicals generated through decarboxylation of primary radical,

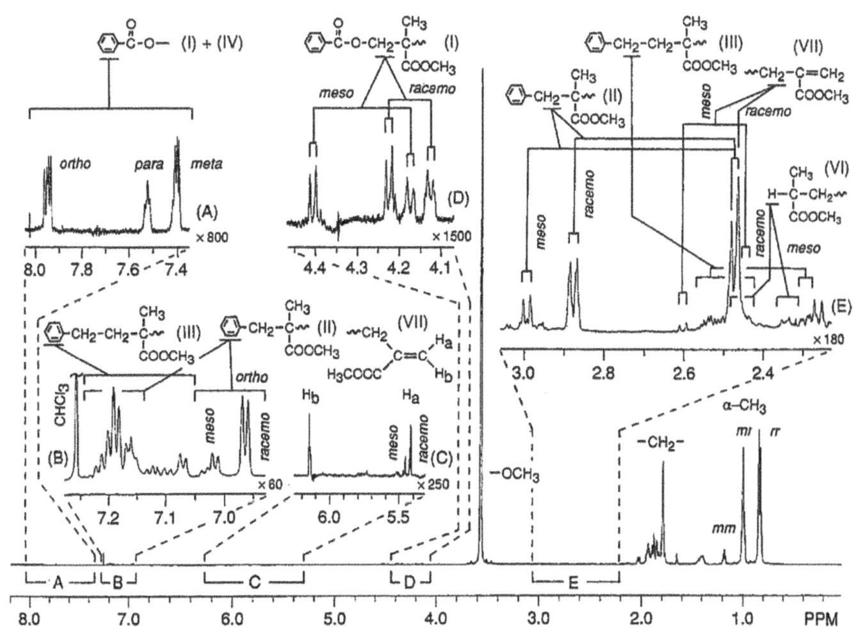

Fig. 4.18 ^{1}H NMR spectrum of PMMA prepared with BPO in toluene at 100 °C [34]

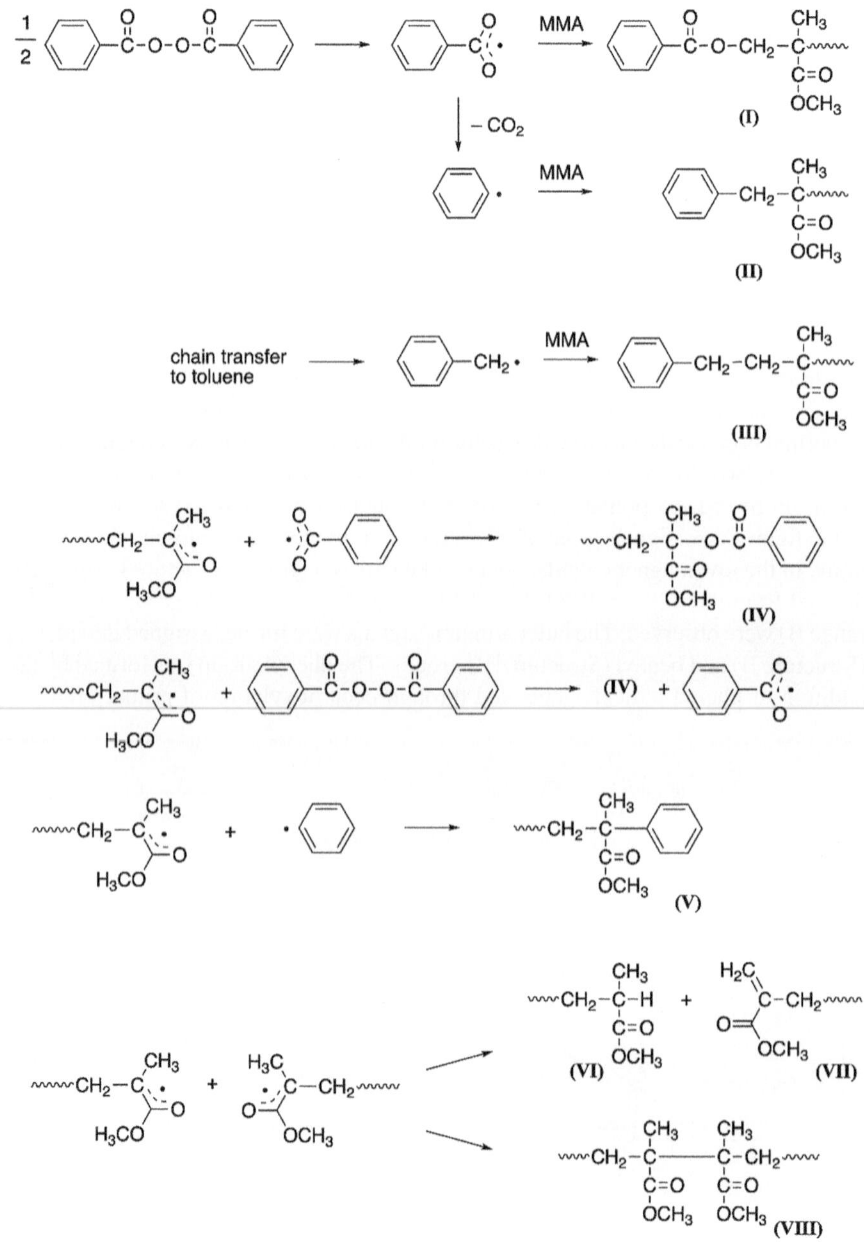

Fig. 4.19 Possible terminal structures formed in radical polymerization of MMA with BPO in toluene [34]

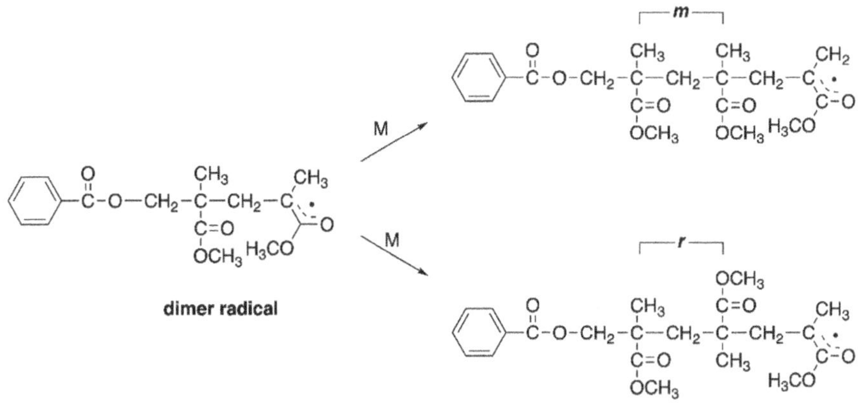

Fig. 4.20 Propagating reaction by MMA-dimer radical determining the first diad tacticity

whereas the benzyl group was formed by the addition of benzyl radicals generated by chain-transfer to toluene used as a solvent (Fig. 4.19). These assignments were made by comparing the ^{1}H NMR spectra of PMMA anionically prepared with C_6H_5MgBr and $C_6H_5CH_2Li$, respectively.

The methylene groups of the first monomeric unit adjacent to the benzoyloxy fragment were observed in range D. The two protons of the methylene groups are nonequivalent due to the adjacent asymmetric carbon, resulting in AB quartet spectral pattern. Furthermore, they show the splitting due to tacticity of the first diad of the polymer. The *m*/*r* ratio determined from the signal intensities was 41/59. This is quite different from the ratio 25/75 determined from the in-chain triad tacticity. This means that the stereoselectivity of the dimer radicals differs from that of the polymer radicals (Fig. 4.20).

The intensity ratio of the benzoyloxy and methylene groups is expected to be 5.0/2.0. However, the observed ratio was 5.8/2.0, suggesting a different type of incorporation of the initiator fragment. The most possible incorporation is attachment to the terminating chain-end (Structure IV) formed by termination reaction between the propagating and primary radicals or chain transfer of the propagating radical to BPO. Otherwise, head addition of benzoyloxy radical to MMA occurred in the initiation process [35].

In the olefinic region (range C), signals of the vinylidene group at the chain end are observed. In the radical polymerization of MMA, disproportionation predominantly occurs more than combination. The disproportionation provides saturated (Structure VI) and unsaturated (Structure VII) chain ends (Fig. 4.19). The signals also show splitting due to tacticity of the terminal diad. The *m*/*r* ratio was found to be 26/74, which is very close to the in-chain tacticity (25/75). This means that two radical termination reaction is independent of the stereochemistry near the propagating chain-ends.

The anionic polymerization of alkyl methacrylates in toluene with a combination of *tert*-C_4H_9Li and bulky aluminum phenoxide gives heterotactic polymers in a living manner [36–39]. Heterotactic polymers comprise an alternating repetition of *m* and *r* diads. As depicted from the stereochemical structure of heterotactic polymer, there exist two kinds of propagating anions, *r*-ended and *m*-ended anions, in this polymerization system (Fig. 4.21). Direct observation of these two types of anions is difficult. However, chain-end analysis of the polymers isolated after quenching was found useful to determine the fraction of the two types of anions [37].

The terminal methine carbon, which formed by reaction with proton, showed principally four split signals in ^{13}C NMR spectra. The splitting reflects four possible triad sequences at the terminating chain-end, ~*mm*, ~*mr*, ~*rm*, and ~*rr*, and were assigned as shown in Fig. 4.22. The assignments were made by comparing the spectra of three kinds of stereoregular polymers prepared *via* living anionic polymerizations [37]. The ultimate diads were fixed when the polymerization was quenched with alcohols. Therefore, the terminal diad reflects the stereoselectivity of the protonation

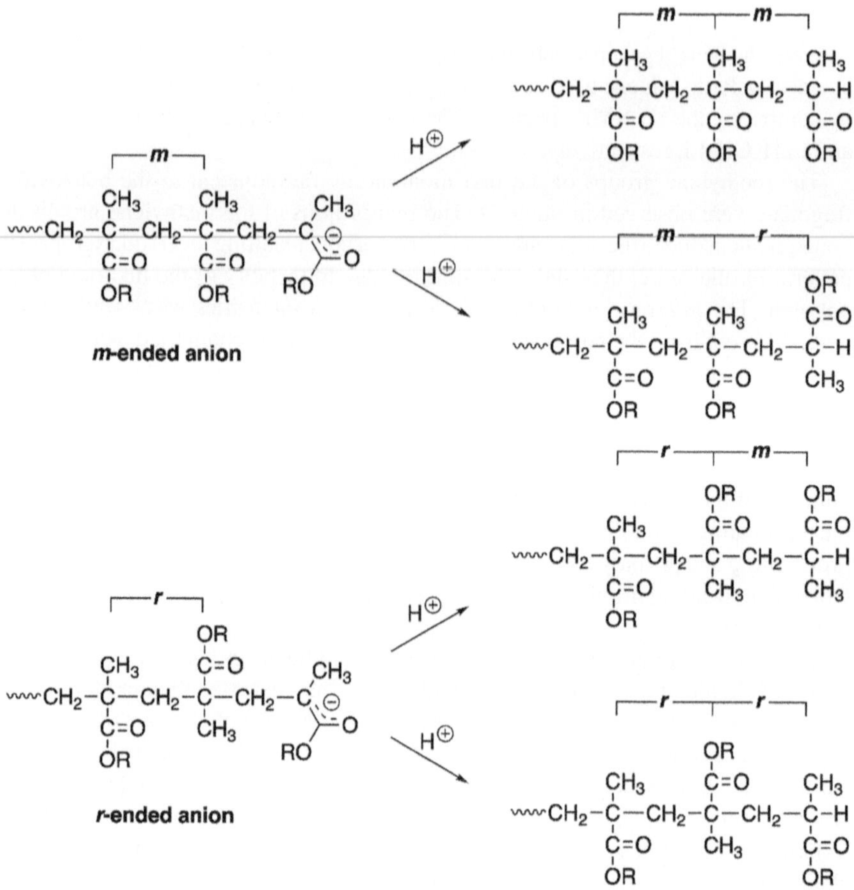

Fig. 4.21 Termination reaction of propagating anions with H$^+$ in heterotactic-specific polymerization

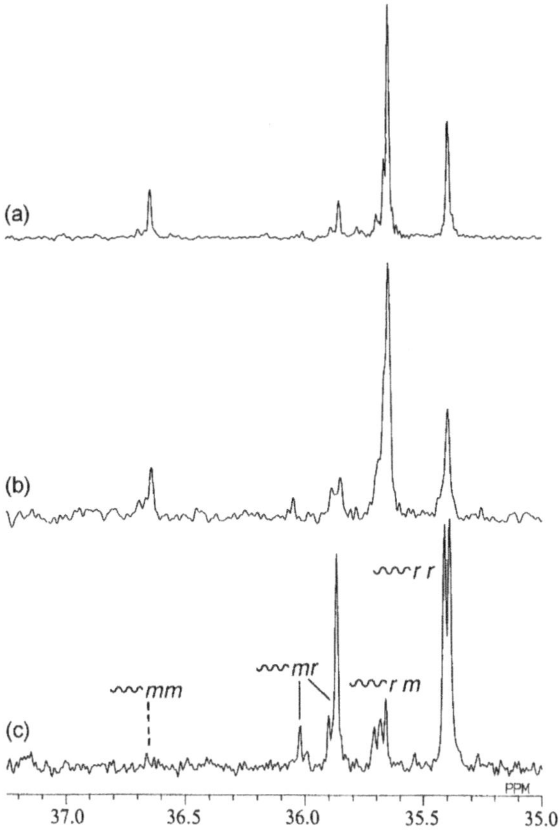

Fig. 4.22 125 MHz ^{13}C NMR spectra of methine carbon at the terminating chain-end of heterotactic poly(ethyl methacrylate)s formed at (**a**) 28%, (**b**) 41%, and (**c**) 100% conversions, respectively [38]

reaction. The *m*/*r* ratio of the second diad from the terminal should correspond to the ratio of the *m*-ended and *r*-ended anions existing in the polymerization system before quenching the reaction; the ratio of $[\sim\sim\sim mM^-]/[\sim\sim\sim rM^-]$ can be estimated from the terminal triad fractions as follows:

$$\frac{[\sim\sim\sim mM^-]}{[\sim\sim\sim rM^-]} = \frac{[\sim mm] + [\sim mr]}{[\sim rm] + [\sim rr]}$$

The ratio was found to be 22/78 for the polymer obtained at 28% conversion. This indicates that the *r*-ended anion is more abundant than the *m*-ended anion, suggesting the higher stability of the *r*-ended anion in this polymerization system. The spectral pattern became complicated with conversion, probably because of the increasing irregularity of the chain-end tacticity at the later stage of the polymerization. Notably, the intensity of the signal of ~*rm* relative to that of ~*rr* was conversed at 100% conversion. This suggests that the stereoselectivity of the protonation reaction depends on the monomer conversion.

4.3 Chromatographic Techniques

4.3.1 HPLC for Characterization of Polymers

High performance liquid chromatography (HPLC) is separation and detection method in which a liquid mobile phase is pressurized by a pump and passed through a column to separate and detect analytes using differences in interactions such as adsorption, distribution, ion exchange, and size exclusion between the stationary phase and the mobile phase. HPLC is widely used for the characterization of polymers as well. Polymers are generally heterogeneous in molecular weight, tacticity, microstructure, chemical composition, chain architecture, and so on. Especially HPLC technique is a very powerful method for characterization of molecular weight (molecular size) and molecular weight distribution. In addition to such analytical purposes, HPLC is also used as a preparative device for separation and purification of natural products and chemically synthesized products [40–45].

4.3.2 Device Configuration of HPLC

Chromatography is a method of separating substances by utilizing the fact that substances are distributed in a certain ratio due to differences in interaction between the stationary phase and the mobile phase that flows in contact with it, and that the ratio varies from substance to substance [40–42]. Figure 4.23 shows the basic device configuration of HPLC.

The mobile phase used in HPLC is called the eluent and plays an important role in separation. Water, aqueous solutions to which various salts are added, and organic solvents such as tetrahydrofuran, methanol, acetonitrile, and hexane are used as the mobile phase, either alone or in mixture. The mobile phase is pumped at a constant flow rate (e.g., 1 mL/min) by a pump driven by a small pulse motor to the column where the components are separated. The flow rate range of the pump is ~1.0 mL/min for analytical use and ~100 mL/min for preparative use, requiring high flow stability and pressure resistance. However, with the recent development of

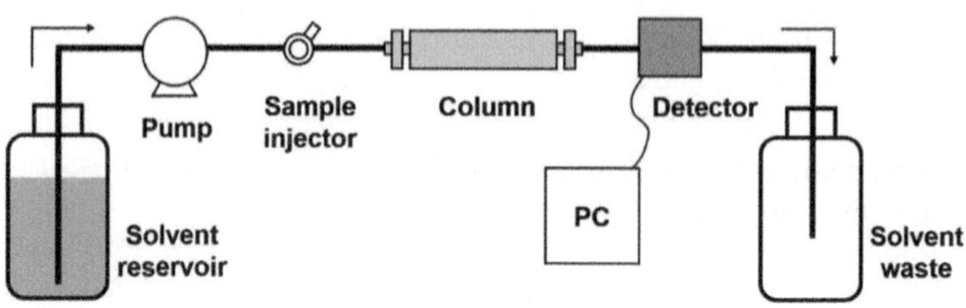

Fig. 4.23 Basic device configuration of HPLC

finer packing materials and downsizing of columns, as described below, pumps with pressure resistance of 100 MPa or higher and pumping stability at low flow rates (μL/min or lower) have been developed.

4.3.3 Application of HPLC to Polymer Analysis: SEC

HPLC has made remarkable progress since its birth in the late 1960s. In addition to advances in hardware and column technology, the wide applicability of HPLC can be attributed to its development. In particular, HPLC has developed mainly in the pharmaceutical, food, and biochemical fields, where components are often thermally unstable or pyrolytically decomposed, therefore they are generally difficult to analyze by GC.

HPLC has made a significant contribution to the pharmaceutical field, where it is widely used for the analysis of impurities in raw materials, active ingredients in products, and metabolic components. Since vitamins are thermally unstable, HPLC analysis plays an active role in product development and quality control. Among the various applications of HPLC, it is especially essential for the analysis of polymers. Among them, size-exclusion chromatography (SEC) or gel-permeation chromatography (GPC) is a typical method of polymer analysis and is used to measure the molecular weight and molecular weight distribution of polymers. The relative molecular weight of polymers can be easily determined by simply dissolving the sample in a solvent (e.g., tetrahydrofuran) and injecting it into the sample. Figure 4.24a shows a SEC chromatograms using standard polymer samples of known molecular weight with narrow molecular weight distribution (e.g., polystyrene), a calibration curve (graph of elution time vs. molecular weight) as shown in Fig. 4.24b is created from the SEC chromatograms, and various average molecular weights (number average molecular weight, weight average molecular weight, etc.) are calculated using software on the data processing unit [40–42].

When there exists a simple relationship between the chain dimension and its molecular weight for chemically homogenous linear polymers, SEC retention time is correlated with the molecular weight of the polymer analyzed. But in case the polymer has nonlinear chain architecture such as branched and cyclic polymers, and the polymer is copolymer and polymer blend (polymer mixture), a simple relationship between chain dimension and molecular weight does not hold, and SEC is not able to separate them according to their molecular weight. Furthermore, SEC is not able to separate polymers efficiently in terms of chemical heterogeneity, such as chemical composition differences of copolymers, microstructure, and tacticity.

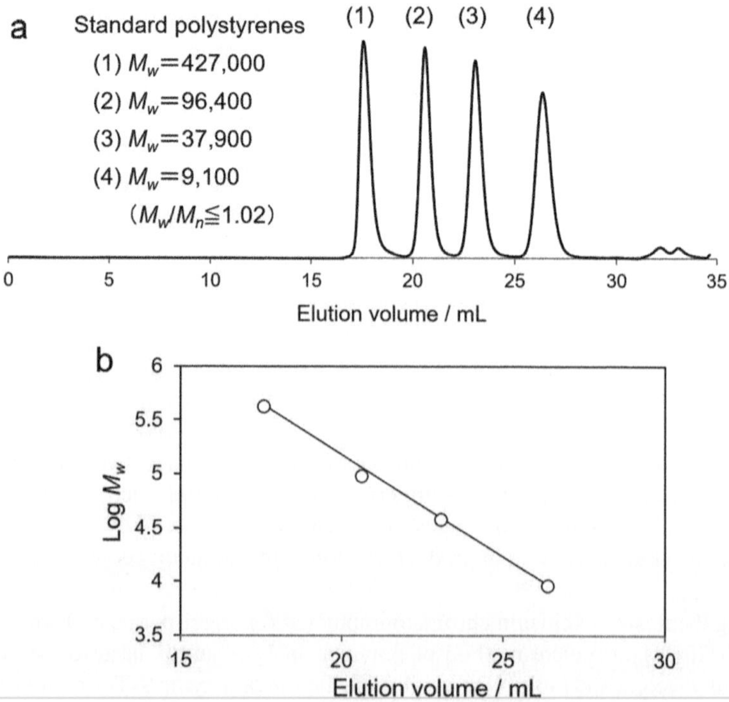

Fig. 4.24 **a** SEC chromatograms for standard polystyrenes, **b** calibration curve created from four standard polystyrene samples

4.3.4 Three Types of HPLC Elution Modes Used for Polymer Analysis: SEC, IC, LCCC

HPLC is widely used for the characterization of polymers, the most frequently used method is SEC as mentioned above, SEC uses a column which does not interact with the polymers, and the polymer molecules are separated by the chain dimension of a polymer in solution. Polymer molecules with larger apparent size (molecular weight) are eluted first as shown in Fig. 4.25a [40–42].

While interaction chromatography (IC) has different separation mechanism, IC uses a column having an interaction between the sample and the column, and the sample is eluted by repeatedly adsorbing and desorbing in the column. Therefore, molecules of the same polarity elute from molecules with fewer adsorption points, i.e., molecules with lower molecular weight are eluted first as shown in Fig. 4.25b [43, 44].

When a sample is eluted by the IC mechanism, solvent gradient interaction chromatography (SGIC) and temperature gradient interaction chromatography (TGIC) are typical methods for separation by gradually changing the solvent composition or temperature as a means of changing the adsorption power. TGIC, in particular, uses a

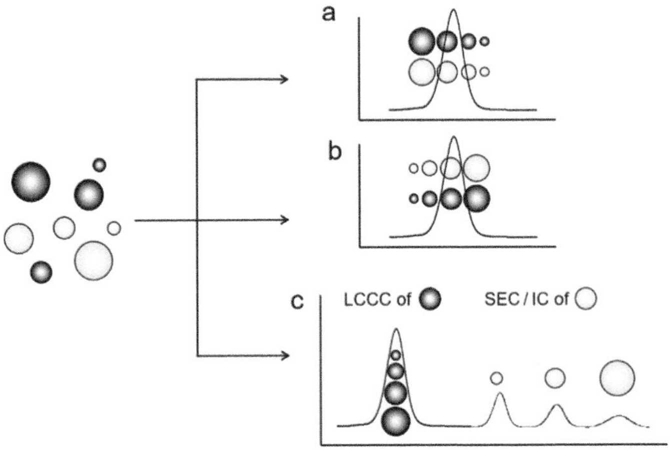

Fig. 4.25 **a** SEC mode, **b** IC mode, **c** LCCC mode

circulating thermostatic chamber having a programmable temperature control mechanism to control the column temperature, enabling high-resolution and reproducible separations.

As mentioned above, SEC and IC mode show exactly opposite elution behaviors. If it is possible to change the polarity of the stationary and mobile phases and the adsorption force of the solute on the stationary phase, it should be possible to control the elution behavior from the SEC mode to the IC mode even using a usual HPLC column. For samples with the same polarity, the SEC mode elutes from high molecular weight samples to low molecular weight ones, while the IC mode elutes from low molecular weight samples to high molecular weight ones. Therefore, if appropriate conditions are selected between the two elution modes, samples of the same type of macromolecule with the same chain architecture will appear in almost the same elution volume, independent of molecular weight. HPLC condition utilizing such chromatographic critical conditions is called liquid chromatography at the chromatographic critical condition (LCCC) [43–45]. For example, it is very useful for separating polymers with different chain architectures, different polymer mixtures, block copolymers, polymers with different stereo-regularities, and even polymers with different terminal functional groups.

4.3.5 Determination of Purity of Cyclic Polymers Using LCCC

Study on cyclic polymers has been a quite interesting subject in polymer science for many years, since they are expected to have quite different properties from those of the corresponding linear ones [46, 47]. Since intermolecular coupling reactions also compete in the cyclization reaction, various kinds of condensation products are formed. From the cyclization reaction product, cyclic polymers are usually isolated and purified by fractionation. So far, the characterization of the obtained cyclic polymers was carried out by SEC, or viscosity measurement measurements, etc. From the facts that the observed values such as elution time in SEC, and intrinsic viscosity were different from those of the corresponding linear ones, the researchers actually assumed there exists considerable amounts of cyclic molecules in the fractionated samples. Recently, proving of the cyclic structures was attempted by NMR analysis of linking points, which gives direct evidence for the cyclic structures, however it cannot be applicable to high molecular weight polymers [48]. However, the presence or absence of linear polymers in the fractionated sample cannot be sufficiently quantitatively analyzed using SEC or other methods. This is because the shrinking factors, which is ratio of radius of gyration of cyclic polymers and that for linear polymers of equal molecular weight, is approximately 0.7, and the separation capability of SEC is not sufficient to completely separate the peaks of the two polymers [47, 49, 50]. However, if, for example, cyclic polymers are measured under LCCC or IC conditions for linear polymers, the elution positions of linear and cyclic molecules will change significantly, allowing complete separation. Figure 4.26 presents an example of finding LCCC and IC conditions for linear polystyrene (PS) using three types of standard polystyrenes (Mw = 10, 40, 100 k) in THF/hexane mixture (42/58, v/v) using bare silica column (5SIL10E, Shodex Co.) at different temperatures from 20 to 50 °C [51].

In the high temperature range of 35–50 °C, the samples were eluted from high molecular weight samples, which elution mode is SEC. On the other hand, in the low temperature range of 20–30 °C, the sample is eluted from low molecular weight

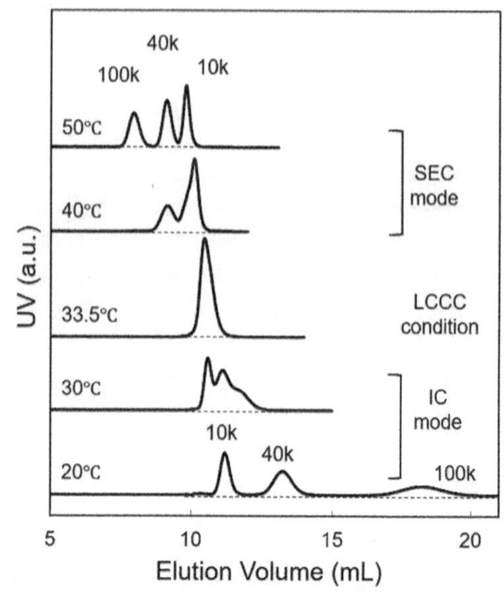

Fig. 4.26 HPLC chromatograms of standard polystyrenes (M_w = 10, 40, and 100 k) at different temperature ranging from 20 to 50 °C

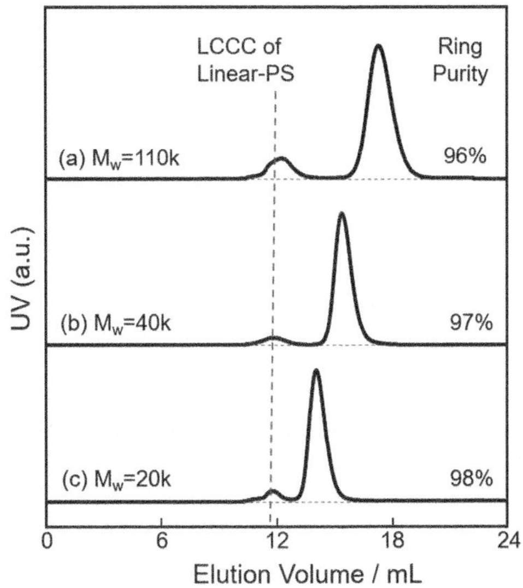

Fig. 4.27 LCCC chromatograms of fractionated cyclic polystyrenes ($M_W = 20, 40$, and 110 k)

samples, indicating that the elution mode is IC. In the intermediate temperature range of 33.5 °C, the three samples with different molecular weights elute at the same elution time, indicating that the elution is not dependent on molecular weight, i.e., the LCCC condition. HPLC charts of synthesized cyclic polystyrene samples with Mw = 20, 40, and 110 k under these LCCC conditions are shown in Fig. 4.27. These three samples were purified by preparative SEC fractionation, and it was quantitatively clear that all cyclic samples contained 2–4% linear molecules [52].

Figure 4.28 shows the LCCC chromatograms of the cyclic polystyrene (Mw = 70 k) which was purified by a combination of SEC and LCCC fractionations, and its linear counterpart. From these comparisons as shown in Fig. 4.28a, the purity of the cyclic molecules was measured to be 99.6%. While the results of SEC measurements of the same two samples are shown in Fig. 4.28b, indicating that the purity of the cyclic molecules cannot be determined quantitatively from peak comparisons because of the overlapping between the two peaks in the SEC analysis [52].

4.3.6 Determination of Purity of Block Copolymers Using TGIC

In the synthesis of block copolymers using sequential addition of monomers by anionic polymerization, deactivated precursor homopolymer (or precursor block polymer) is formed as a byproduct of the reaction between the living block and impurities (proton donor, etc.) contained in the added monomer. Therefore, except in an ideal state where the monomers are perfectly purified, the synthesized block copolymers usually contain precursor homopolymers (or precursor block polymers). Since

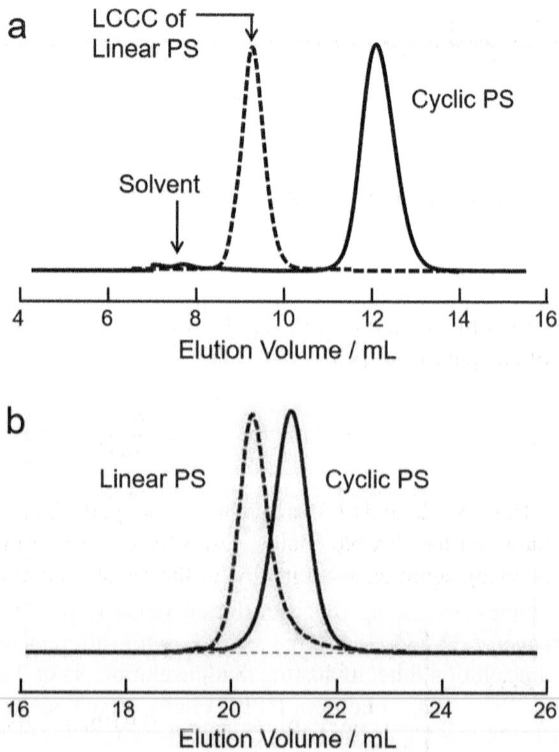

Fig. 4.28 Comparison between **a** LCCC and **b** SEC chromatograms of LCCC-purified cyclic polystyrene and the linear precursor

the contamination of these homopolymers, etc., often has a considerable adverse effect on the physical properties of the block copolymers, it is important to estimate the impurity content, or block copolymer purity. In addition to LCCC, IC is also used for the analysis of block copolymers.

Block copolymers that have been analyzed by HPLC include polystyrene-polyisoprene (PS-PI) [53], polystyrene-polybutadiene (PS-PB) [54], and polystyrene-polymethylmethacrylate (PS-PMMA) [55]. These block polymer systems have found their separation conditions by using silica gel columns and polarity-controlled solvent mixtures such as (CH_2Cl_2/CH_3CN). However, for polystyrene-poly(2-vinylpyridine) (SP) block copolymer system, separation conditions have not been found due to the strong adsorption of P2VP on silica gel columns. Since our research group has been using P2VP in various block polymer systems, we have been searching for HPLC separation conditions for block copolymer

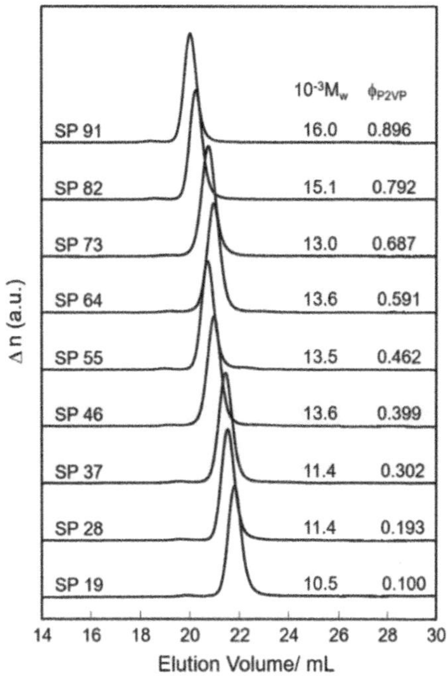

Fig. 4.29 SEC chromatograms of a series of SP diblock copolymers

systems containing P2VP. As a result of our investigation, we were able to find the separation conditions [56]. The column was NH_2-modified silica gel (Hypersil APS-1, 100Åpore, 150 mm × 4.6mm), the solvent was THF/CH_3CN (53/47v/v), to which 0.005M LiCl salt was added and the temperature was varied between 5 and 55 °C. TGIC was used to confirm that P2VP could be separated. Figure 4.29 shows the SEC chromatograms for nine samples of PS-P2VP diblock copolymers with a molecular weight of about 13×10^4 and volume fractions of P2VP (ϕ_{P2VP}) varying from 0.1 to 0.9 in 0.1 increments. From these chromatograms, it is difficult to analyze whether or not the precursor homopolymer is present in these samples. On the other hand, Fig. 4.30 shows the TGIC measurement results of these 9 samples. It can be seen that the polystyrene precursor elutes in about 2.5 minutes, whereas the block copolymer containing P2VP elutes slower than polystyrene. Furthermore, samples with higher P2VP content elute more slowly, indicating a clear separation by chemical composition. In addition, since almost no polystyrene precursor peaks appeared at the elution position of about 2.5 min for any of the block copolymers (less than 1%), it was confirmed that the purity of the block copolymers was very high (greater than 99%). Thc samc analysis of the P2VP-PS-P2VP triblock copolymer also confirmed the high purity (>99%) of the samples.

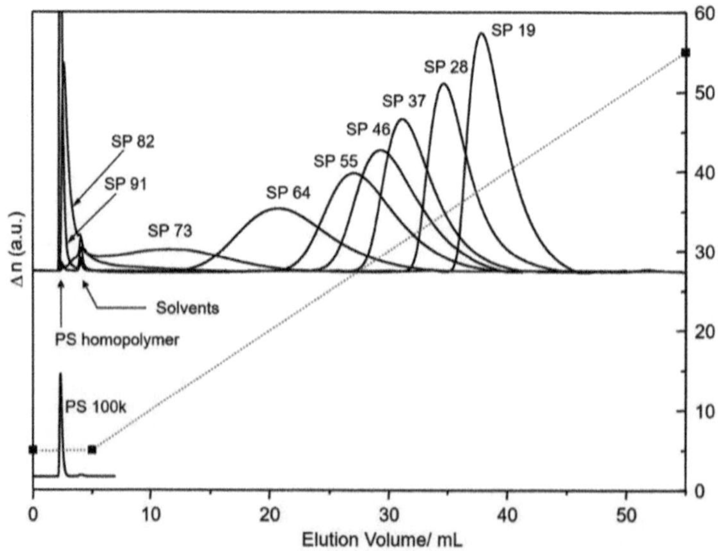

Fig. 4.30 TGIC chromatograms of a series of SP diblock copolymers

References

1. Bovey FA, Tiers GVD (1960) Polymer NSR spectroscopy. II. The high resolution spectra of methyl methacrylate polymers prepared with free radical and anionic initiators. Journal of Polymer Science 44 (143):173–182. https://doi.org/10.1002/pol.1960.1204414315.
2. Nishioka A, Watanabe H, Yamaguchi I, Shimizu H (1960) High resolution NMR spectra of isotactic and syndiotactic polymethyl methacrylate in chloroform solution. Journal of Polymer Science 45 (145):232–234. https://doi.org/10.1002/pol.1960.1204514526.
3. Johnsen U, Tessmar K (1960) Hochaufgelöste kernresonanzspektren von eutaktischen polymethylmethacrylaten. Kolloid-Z 168 (2):160–161. https://doi.org/10.1007/bf01507429.
4. Fellows CM, Hellwich K-H, Meille SV, Moad G, Nakano T, Vert M (2020) Definitions and notations relating to tactic polymers (IUPAC Recommendations 2020). Pure and Applied Chemistry 92 (11):1769–1779. https://doi.org/10.1515/pac-2019-0409.
5. Bovey FA (1972) High resolution NMR of macromolecules. Academic Press, New York.
6. Hatada K, Kitayama T (2004) NMR Spectroscopy of polymers. Laboratory Manual Series in Polymer Science. Springer, Berlin. https://doi.org/10.1007/978-3-662-08982-8.
7. Hirano T, Miyamoto Y, Amano S, Tatsumi K, Anmoto T, Kimura H, Yoshida K, Oshimura M, Ute K (2014) Hydrogen-bond-assisted isotactic-specific radical polymerization of *N*-vinyl-2-pyrrolidone with tartrate additives in toluene at low temperatures: high-resolution ^{1}H NMR analysis. RSC Advances 4 (95):53079–53089. https://doi.org/10.1039/C4RA08743G.
8. Majumdar RN, Harwood HJ (1981) Determination of the microstructure of poly(methyl α-fluoroacrylate) by NMR spectroscopy. Polymer Bulletin 4 (7):391–397. https://doi.org/10.1007/bf00254970.
9. Victor MW, Saffariannour M, Reynolds JR (1994) Structural Characterization of Poly(α-Fluoroacrylonitrile) and Poly(Ethyl α-Fluoroacrylate). Journal of Macromolecular Science, Part A 31 (6):721–736. https://doi.org/10.1080/10601329409349751.
10. Hirano T, Furutani T, Oshimura M, Ute K (2012) Syndiotactic- and heterotactic-specific radical polymerizations of *N*-*n*-propyl-α-fluoroacrylamide and phase-transition behaviors of aqueous

solutions of poly(*N*-*n*-propyl-α-fluoroacrylamide). Journal of Polymer Science Part A: Polymer Chemistry 50 (12):2471–2483. https://doi.org/10.1002/pola.26024.
11. Hirano T, Yamaguchi H, Kitayama T, Hatada K (1998) Heterotactic Polymerization of Methacrylates Having C-3 Ester Group. Polymer Journal 30 (9):767–769. https://doi.org/10.1295/polymj.30.767.
12. Randall JC (1978) A ^{13}C NMR Determination of the Comonomer Sequence Distributions in Propylene-Butene-1 Copolymers. Macromolecules 11 (3):592–597. https://doi.org/10.1021/ma60063a032.
13. Harwood HJ, Ritchey WM (1964) The characterization of sequence distribution in copolymers. Journal of Polymer Science Part B: Polymer Letters 2 (6):601–607. https://doi.org/10.1002/pol.1964.110020607.
14. Coleman BD, Fox TG (1963) Multistate Mechanism for Homogeneous Ionic Polymerization. I. The Diastereosequence Distribution. The Journal of Chemical Physics 38 (5):1065–1075. https://doi.org/10.1063/1.1733803.
15. Coleman BD, Fox TG (1963) General theory of stationary random sequences with applications to the tacticity of polymers. Journal of Polymer Science Part A: General Papers 1 (10):3183–3197. https://doi.org/10.1002/pol.1963.100011013.
16. Yamadera R, Murano M (1967) The determination of randomness in copolyesters by high resolution nuclear magnetic resonance. Journal of Polymer Science Part A-1: Polymer Chemistry 5 (9):2259–2268. https://doi.org/10.1002/pol.1967.150050905.
17. Hirano T, Li M, Maeda K, Oshimura M, Ute K (2020) Unusually large hysteresis in temperature-responsive phase-transition behavior of aqueous solutions of isotactic copolymers comprising *N*-ethylacrylamide and *N*-isopropylacrylamide. Polymer 198:122530. https://doi.org/10.1016/j.polymer.2020.122530.
18. Nishiura T, Kitayama T, Hatada K (2000) Carbon-13 NMR Spectra of Stereoregular Copolymers of Methyl and Butyl Methacrylates. International Journal of Polymer Analysis and Characterization 5 (4-6):401–413. https://doi.org/10.1080/10236660008034635.
19. Momose H, Hattori K, Hirano T, Ute K (2009) Multivariate analysis of ^{13}C NMR spectra of methacrylate copolymers and homopolymer blends. Polymer 50 (15):3819–3821. https://doi.org/10.1016/j.polymer.2009.05.048.
20. Momose H, Maeda T, Hattori K, Hirano T, Ute K (2012) Statistical determination of chemical composition and monomer sequence distribution of poly(methyl methacrylate-*co*-*tert*-butyl methacrylate)s by multivariate analysis of ^{13}C NMR spectra. Polymer Journal 44 (8):808–814. https://doi.org/10.1038/pj.2012.110.
21. Hirano T, Kamiike R, Hsu Y, Momose H, Ute K (2016) Multivariate analysis of ^{13}C NMR spectra of branched copolymers prepared by initiator-fragment incorporation radical copolymerization of ethylene glycol dimethacrylate and *tert*-butyl methacrylate. Polymer Journal 48 (7):793–800. https://doi.org/10.1038/pj.2016.16.
22. Hirano T, Kamiike R, Yuki T, Matsumoto D, Ute K (2022) Determination of monomer reactivity ratios from a single sample using multivariate analysis of the 1H NMR spectra of poly[(methyl methacrylate)-co-(benzyl methacrylate)]. Polymer Journal 54 (5):623–631. https://doi.org/10.1038/s41428-022-00618-y.
23. Fineman M, Ross SD (1950) Linear method for determining monomer reactivity ratios in copolymerization. Journal of Polymer Science 5 (2):259–262. https://doi.org/10.1002/pol.1950.120050210.
24. Kelen T, Tüdõs F (1975) Analysis of the Linear Methods for Determining Copolymerization Reactivity Ratios. I. A New Improved Linear Graphic Method. Journal of Macromolecular Science: Part A - Chemistry 9 (1):1–27. https://doi.org/10.1080/00222337508068644.
25. Tidwell PW, Mortimer GA (1965) An improved method of calculating copolymerization reactivity ratios. Journal of Polymer Science Part A: General Papers 3 (1):369–387. https://doi.org/10.1002/pol.1965.100030137.
26. Hatada K, Kitayama T, Terawaki Y, Sato H, Chujo R, Tanaka Y, Kitamaru R, Ando I, Hikichi K, Horii F (1995) NMR Measurement of Identical Polymer Samples by Round Robin Method IV.

Analysis of Composition and Monomer Sequence Distribution in Poly(methyl methacrylate-*co*-acrylonitrile) Leading to Determinations of Monomer Reactivity Ratios. Polymer Journal 27 (11):1104–1112. https://doi.org/10.1295/polymj.27.1104.

27. Chujo R, Ubara H, Nishioka A (1972) Determination of Monomer Reactivity Ratios in Copolymerization from a Single Sample and Its Application to the Acrylonitrile-Methyl Methacrylate System. Polymer Journal 3 (6):670–674. https://doi.org/10.1295/polymj.3.670.
28. Gerken TA, Ritchey WM (1978) Determination of the monomer sequencing and microstructure of acrylonitrile copolymers by ^{13}-C NMR. I. Acrylonitrile-methyl methacrylate and acrylonitrile-ethyl methacrylate copolymers. J Appl Polym Sci: Appl Polym Symp 34 (Polym. Anal.):17–34.
29. Johnsen VU, Kolbe K (1968) Bestimmung der Copolymerisationsparameter und Depolymerisationskonstanten von Methylmethacrylat-Acrylnitril-Copolymeren durch Protonenspinresonanz. Die Makromolekulare Chemie 116 (1):173–189. https://doi.org/10.1002/macp.1968.021160118.
30. Kowalski BR, Bender CF (1973) Pattern recognition. II. Linear and nonlinear methods for displaying chemical data. Journal of the American Chemical Society 95 (3):686–693. https://doi.org/10.1021/ja00784a007.
31. Kowalski BR, Bender CF (1972) Pattern recognition. Powerful approach to interpreting chemical data. Journal of the American Chemical Society 94 (16):5632–5639. https://doi.org/10.1021/ja00771a016.
32. Wold S (1974) Spline Functions in Data Analysis. Technometrics 16 (1):1–11. https://doi.org/10.1080/00401706.1974.10489142.
33. Ito K, Yamashita Y (1965) Copolymer composition and microstructure. Journal of Polymer Science Part A: General Papers 3 (6):2165–2187. https://doi.org/10.1002/pol.1965.100030607.
34. Hatada K, Kitayama T, Ute K, Terawaki Y, Yanagida T (1997) End-Group Analysis of Poly(methyl methacrylate) Prepared with Benzoyl Peroxide by 750 MHz High-Resolution ^{1}H NMR Spectroscopy. Macromolecules 30 (22):6754–6759. https://doi.org/10.1021/ma970672i.
35. Moad G, Solomon DH (1995) The chemistry of free radical polymerization. Pergamon, Oxford, OX, U.K.
36. Kitayama T, Zhang Y, Hatada K (1994) Preparation of Highly Heterotactic Polymethacrylate with Narrow Molecular Weight Distribution by *t*-Butyllithium/Bis(2,6-di-*t*-butylphenoxy)methylaluminum in Toluene. Polymer Journal 26 (7):868–872. https://doi.org/10.1295/polymj.26.868.
37. Kitayama T, Hirano T, Hatada K (1996) Chain-End Stereostructure of Heterotactic Poly(ethyl methacrylate) Prepared with *t*-Butyllithium/Bis(2,6-di-*t*-butylphenoxy)methylaluminum as Studied by ^{13}C NMR Spectroscopy. Polymer Journal 28 (1):61–67. https://doi.org/10.1295/polymj.28.61.
38. Kitayama T, Hirano T, Hatada K (1996) Remarkable Change of Stereochemistry near the Terminating Chain-End of Heterotactic Poly(ethyl methacrylate) with Conversion. Importance in Mechanistic Study of Stereospecific Polymerization with the Aid of Chain-End Analysis. Polymer Journal 28 (12):1110–1111. https://doi.org/10.1295/polymj.28.1110.
39. Hirano T, Kitayama T, Cao J, Hatada K (2000) ^{13}C NMR Analysis of Uniform Heterotactic Oligo(allyl methacrylate) Aiming at Understanding Even-Odd Alternation in Distribution of Degree of Polymerization. Polymer Journal 32 (11):961–969. https://doi.org/10.1295/polymj.32.961.
40. Yau WW, Kirkland JJ, Bly DD (1979) Modern size-exclusion liquid chromatography: practice of gel permeation and gel filteration chromatography. Wiley, New York.
41. Potschka M, Dublin PL (eds) (1996) Strategies in size exclusion chromatography. American Chemical Society, Washington DC.
42. Mori S, Barth HG (1999) Size exclusion chromatography. Springer, Berlin, Heidelberg, New York.
43. Patsh H (1997) Analysis of Complex Polymers by Interaction Chromatography. Adv Polym Sci 128: 1–45. https://doi.org/10.1007/3-540-61218-1

44. Chang C (2003) Recent Advances in Liquid Chromatography Analysis of Synthetic Polymers. Adv Polym Sci 163: 1–60. https://doi.org/10.1007/b11050
45. Marko T, Hunkeler D (2003) Liquid Chromatography under Critical and Limiting Conditions: A Survey of Experimental Systems for Synthetic Polymers. Adv Polym Sci 163: 62–136. https://doi.org/10.1007/b11053
46. Hild G, Kohler A, Rempp P (1980) Synthesis of ring-shaped macromolecules. Eur Polym J 16: 525–527. https://doi.org/10.1016/0014-3057(80)90136-6
47. Roovers J, Topoloski P (1983) Synthesis of High Molecular Weight Ring Polystyrenes. Macromolecules 16: 843–849. https://doi.org/10.1021/ma00240a002
48. Kubo M, Hayashi T, Kobayashi H, Tsuboi K, Itoh T (1997) Synthesis of α-Carboxyl, ω-Amino Heterodifunctional Polystyrene and Its Intramolecular Cyclization. Macromolecules 30: 2805–2807. https://doi.org/10.1021/ma961850v
49. Zimm BH, Stockmayer WH (1949) The Dimensions of Chain Molecules Containing Branches and Rings. J Chem Phys 17: 1301–1314. https://doi.org/10.1063/1.1747157
50. Takano A, Kushida K, Aoki K, Masuoka K, Hayashida K, Cho D, Kawaguchi D, Matsushita Y (2007) HPLC Characterization of Cyclization Reaction Product Obtained by End-to-End Ring Closure Reaction of a Telechelic Polystyrene. Macromolecules 40: 679–681. https://doi.org/10.1021/ma062496w
51. Doi Y, Ohta Y, Nakamura M, Takano A, Takahashi Y, Matsushita Y (2013) Precise synthesis and characterization of tadpole-shaped polystyrenes with high purity. Macromolecules 46: 1075–1081. https://doi.org/10.1021/ma302511j
52. Cho D, Masuoka K, Koguchi K, Asari T, Takano A, Matsushita M (2005) Preparation and Characterization of Cyclic Polystyrenes. Polym J 37: 506–511. https://doi.org/10.1295/polymj.37.506
53. Park I, Park S, Cho D, Chang T, Kim E, Lee K, Kim YJ (2003) Effect of Block Copolymer Chain Architecture on Chromatographic Retention. Macromolecules 36: 8539–8543. https://doi.org/10.1021/ma035033o
54. Lee W, Cho D, Chang T, Hanley KJ, Lodge TP (2001) Characterization of polystyrene-b-polyisoprene diblock copolymers by liquid chromatography at the chromatographic critical condition. Macromolecules 34: 2353–2358. https://doi.org/10.1021/ma001727a
55. Zimina TM, Fell AF, Castledine JB (1992) Diode-array detection with spectral suppression in size exclusion and adsorption chromatography for rapid evaluation of chemical composition in block copolymers of styrene and methyl methacrylate. Polymer 33: 4129–4135. https://doi.org/10.1016/0032-3861(92)90616-5
56. Cho D, Noro A, Takano A, Matsushita Y (2005) TGIC Separation of PS-b-P2VP Diblock and P2VP-b-PS-b-P2VP Triblock Copolymers According to Chemical Composition. Macromolecules 38: 3033–3036. https://doi.org/10.1021/ma0474877

Index

B
Bernoullian statistics, 78
Berry plot, 40

C
Chain-end group, 93
Chemical composition, 85
Combination, 95
Contour length, 31

D
Disproportionation, 95

E
Electro-spray ionization, 22
Excluded-volume effect, 54
Excluded-volume strength, 54

F
Flory-Fox factor, 19

G
Gel-Permeation Chromatography (GPC), 19, 99
Guinier plot, 40

H
High Performance Liquid Chromatography (HPLC), 98
Holtzer plot, 42
Huggins constant, 18
Hydrodynamic radius, 51

I
Interaction Chromatography (IC), 100
Intrinsic viscosity, 17

K
Kratky plot, 42

L
Light scattering, 13
Liquid Chromatography at the Chromatographic Critical Condition (LCCC), 101
Low-Angle Light Scattering (LALS), 22

M
Magnitude of scattering vector, 15
Mark-Houwink-Sakurada equation, 18
Markovian statistics, 80
Matrix-aided laser desorption ionization, 22
Mean-square radius of gyration, 15, 31
Molecular weight per unit contour length, 32
Monomer reactivity ratio, 86
Monomer sequence, 84
Multi-Angle Light Scattering (MALS), 22

N
Nuclear Magnetic Resonance (NMR), 71

Y. Nakamura et al., *Characterization of Polymers in Solution*,
https://doi.org/10.1007/978-981-96-6677-5

Number-average molecular weight, 6

O
Osmotic pressure, 7

P
Particle scattering function, 34
Persistence length, 31
Polydispersity index, 7
Polymerization mechanism, 93

R
Radius expansion factor, 54
Rayleigh ratio, 13, 33

S
Scattering vector, 34
SEC-MALS method, 22, 47
Second virial coefficients, 36
Sedimentation coefficient, 48
Sedimentation equilibrium, 11
Size-Exclusion Chromatography (SEC), 19, 99
Small-Angle Neutron Scattering (SANS), 37
Small-Angle X-ray Scattering (SAXS), 37
Solvent Gradient Interaction Chromatography (SGIC), 100
Stereochemistry, 72
Stiffness parameter, 31

T
Tacticity, 72
Temperature Gradient Interaction Chromatography (TGIC), 100
Third virial coefficients, 36
Time correlation function, 48

V
Vapor pressure osmometry, 10
Virial expansion, 8

W
Weight-average molecular weight, 6

Z
z-average molecular weight, 6
z-average particle scattering function, 36
Zimm plot, 40

GPSR Compliance

The European Union's (EU) General Product Safety Regulation (GPSR) is a set of rules that requires consumer products to be safe and our obligations to ensure this.

If you have any concerns about our products, you can contact us on ProductSafety@springernature.com

In case Publisher is established outside the EU, the EU authorized representative is:

Springer Nature Customer Service Center GmbH
Europaplatz 3
69115 Heidelberg, Germany

Batch number: 08743412

Printed by Printforce, the Netherlands